RULE ONE

Search your feelings

LUKE SKYWALKER: 'No . . . that's not true. That's impossible!'

DARTH VADER: 'Search your feelings, you know it to be true!'

—*The Empire Strikes Back* (1980)[1]

Abraham Bredius was nobody's fool. An art critic and collector, he was the world's leading scholar on Dutch painters, and particularly the seventeenth-century master Johannes Vermeer. As a young man in the 1880s, Bredius had made his name by spotting works wrongly credited to Vermeer. At the age of eighty-two, in 1937, he was enjoying something of a retirement swansong. He had just published a highly respected book in which he had identified two hundred fakes or imitations of Rembrandt.[2]

It was at this moment in Bredius's life that a charming lawyer named Gerard Boon paid a visit to his Monaco villa. Boon wanted to ask Bredius's opinion of a newly rediscovered work, *Christ at Emmaus*, thought to have been painted by Vermeer himself. The exacting old man was spellbound. He sent Boon away with his verdict: *Emmaus* was not only a Vermeer, it was the Dutch master's finest work.

'We have here – I am inclined to say – the masterpiece of Johannes Vermeer of Delft,' wrote Bredius in a magazine article shortly after. 'Quite different from all his other paintings and yet every inch a Vermeer.

'When this masterpiece was shown to me I had difficulty controlling my emotion,' he added, noting reverently that the work was *ongerept* – Dutch for virginally pure and untouched. It was an ironic choice of words: *Emmaus* could hardly have been more corrupt. It was a rotten fraud of a painting, stiffly applied to an old canvas just a few months before Bredius caught sight of it, and hardened with Bakelite.

Yet this crude trickery not only caught out Bredius, but the entire Dutch art world. *Christ at Emmaus* soon sold for 520,000 guilders to the Boijmans Museum in Rotterdam. Compared to the wages of the time that is the equivalent of about £10 million today. Bredius himself contributed to help the museum buy the picture.

Emmaus became the centrepiece of the Boijmans Museum, drawing admiring crowds and rave reviews. Several other paintings in a similar style soon emerged. Once the first forgery had been accepted as a Vermeer, it was easier to pass off these other fakes. They didn't fool everyone, but like *Emmaus* they fooled the people who mattered. Critics certified them; museums exhibited them; collectors paid vast sums for them – a total of more than £100 million in today's money. In financial terms alone, this was a monumental fraud.

But there was more. The Dutch art world revered Vermeer as one of the greatest painters who ever lived. Painting mostly in the 1660s, he had been rediscovered only in the late 1800s.

Fewer than forty of his works survive. The apparent emergence of half a dozen Vermeers in just a few years was a major cultural event.

It was also an event that should have strained credulity. But it did not. Why?

Don't look to the paintings themselves for an answer. If you compare a genuine Vermeer to the first forgery, *Emmaus*, it is hard to understand how anyone was fooled – let alone anyone as discerning as Abraham Bredius.

Vermeer was a true master. His most famous work is *Girl With a Pearl Earring*, a luminous portrait of a young woman: seductive, innocent, adoring and anxious all at once. The painting inspired a novel, and a movie starring Scarlett Johansson as the unnamed girl. In *The Milkmaid*, a simple scene of domesticity is lifted by details such as the rendering of a copper pot, and a display of fresh-baked bread that looks good enough to grab out of the painting. Then there's *Woman Reading a Letter.* She stands in the soft light of an unseen window. Is she, perhaps, pregnant? We see her in profile as she holds the letter close to her chest, eyes cast down as she reads. There's a dramatic stillness about the image – we feel that she's holding her breath as she scans the letter for news; we hold our breath too. A masterpiece.

And *Christ at Emmaus*? It's a static, awkward image by comparison. Rather than seeming to be an inferior imitation of Vermeer, it doesn't look like a Vermeer at all. It's not a terrible painting, but it's not a brilliant one either. Set alongside Vermeer's works it seems dour and clumsy. And yet it, and several others, fooled the world – and might continue to fool the world to this day, had not the forger been caught out by a combination of recklessness and bad luck.

In May 1945, with the war in Europe at an end, two officers from the Allied Art Commission knocked on the door of 321

Keizersgracht, one of Amsterdam's most exclusive addresses. They were met by a charismatic little man called Han van Meegeren. The young van Meegeren had enjoyed some brief success as an artist. In middle age, as his jowls had loosened and his hair had silvered, he had grown rich as an art dealer.

But perhaps he had been dealing art with the wrong people, because the officers came with a serious charge: that van Meegeren had sold Johannes Vermeer's newly discovered masterpiece, *The Woman Taken in Adultery*, to a German Nazi. And not just any Nazi, but Hitler's right-hand man, Hermann Göring.

Van Meegeren was arrested and charged with treason. He responded with furious denials, trying to bluster his way to freedom. His forceful, fast-talking manner was usually enough to get him out of a sticky situation. Not this time. A few days into his incarceration, he cracked. He confessed not to treason but to a crime that caused astonishment across the Netherlands and the art world as a whole.

'Fools!' he sneered. 'You think I sold a priceless Vermeer to Göring? There was no Vermeer! I painted it myself.'[3]

Van Meegeren admitted painting not only the work that had been found in Nazi hands, but *Christ at Emmaus* and several other supposed Vermeers. The fraud had unravelled not because anyone spotted these flawed forgeries, but because the forger himself confessed. And why wouldn't he? Selling an irreplaceable Vermeer masterpiece to the Nazis would have been a hanging offence, whereas selling a forgery to Hermann Göring wasn't just forgivable, it was admirable.

But the question remains: how could a man as expert as Abraham Bredius have been fooled by so crass a forgery? And why begin a book about statistics with a tale that has nothing at all to do with numbers?

The answer to both questions is the same: when it comes

to interpreting the world around us, we need to realise that our feelings can trump our expertise. When Bredius wrote 'I had difficulty controlling my emotion', he was, alas, correct. Nobody had more skill or knowledge than Bredius, but van Meegeren understood how to turn Bredius's skill and knowledge into a disadvantage.

Working out how van Meegeren fooled Bredius teaches us much more than a footnote in the history of art; it explains why we buy things we don't need, fall for the wrong kind of romantic partner, and vote for politicians who betray our trust. In particular, it explains why so often we buy into statistical claims that even a moment's thought would tell us cannot be true.

Van Meegeren wasn't an artistic genius, but he intuitively understood something about human nature. Sometimes, we want to be fooled.

We'll return to the cause of Abraham Bredius's error in a short while. For now, it's enough to understand that his deep knowledge of Vermeer's paintings proved to be a liability rather than an asset. When he saw *Christ at Emmaus*, Bredius was undone by his emotional response. The same trap lies in wait for any of us.

The aim of this book is to help you be wiser about statistics. That means I also need to help you be wiser about yourself. All the statistical expertise in the world will not prevent you believing claims you shouldn't believe and dismissing facts you shouldn't dismiss. That expertise needs to be complemented by control of your own emotional reactions to the statistical claims you see.

In some cases there's no emotional reaction to worry about. Let's say I tell you that Mars is more than 50 million kilometres, or 30 million miles, away from the

Earth. Very few people have a passionately held belief about that claim, so you can start asking sensible questions immediately.

For example: is 30 million miles a long way? (Sort of. It's more than a hundred times further than the distance between Earth and the moon. Other planets are a lot further away, though.) Hang on, isn't Mars in a totally different orbit? Doesn't that mean the distance between the Earth and Mars varies all the time? (Indeed it does. The minimum distance between the two planets is a bit more than 30 million miles, but sometimes Mars is more than 200 million miles away.) Because there is no emotional response to the claim to trip you up, you can jump straight to trying to understand and evaluate it.

It's much more challenging when emotional reactions are involved, as we've seen with smokers and cancer statistics. Psychologist Ziva Kunda found the same effect in the lab, when she showed experimental subjects an article laying out the evidence that coffee or other sources of caffeine could increase the risk to women of developing breast cysts. Most people found the article pretty convincing. Women who drank a lot of coffee did not.[4]

We often find ways to dismiss evidence that we don't like. And the opposite is true, too: when evidence seems to support our preconceptions, we are less likely to look too closely for flaws.

The more extreme the emotional reaction, the harder it is to think straight. What if your doctor told you that you had a rare form of cancer, and advised you not to look it up? What if you ignored that advice, consulted the scientific literature, and discovered that the average survival time was just eight months?

Exactly that situation confronted Stephen Jay Gould, a

palaeontologist and wonderful science writer, at the age of forty. 'I sat stunned for about fifteen minutes . . .' he wrote in an essay that has become famous. You can well imagine his emotions. *Eight months to live. Eight months to live. Eight months to live.* 'Then my mind started to work again, thank goodness.'[5]

Once his mind did start to work, Gould realised that his situation might not be so desperate. The eight months wasn't an upper limit; it was the median average, which means that half of sufferers live longer than that. Some, possibly, live a great deal longer. Gould had a good chance: he was fairly young; his cancer had been spotted early; he'd get good treatment.

Gould's doctor was being kind in trying to steer him away from the literature, and many of us will go to some lengths to avoid hearing information we suspect we might not like. In another experiment, students had a blood sample taken and were then shown a frightening presentation about the dangers of herpes; they were then told that their blood sample would be tested for the herpes virus. Herpes can't be cured, but it can be managed, and there are precautions a person can take to prevent transmitting the virus to sexual partners – so it would be useful to know whether or not you have herpes. Nevertheless, a significant minority, one in five, not only preferred not to know whether they were infected but were willing to pay good money to have their blood sample discarded instead. They told researchers they simply didn't want to face the anxiety.[6]

Behavioural economists call this 'the ostrich effect'. For example, when stock markets are falling, people are less likely to log in to check their investment accounts online.[7] That makes no sense. If you use information about share prices to inform your investment strategy, you should be just as keen to get it in bad times as good. If you don't, there's little reason to log in at all – so why check your account so frequently when the market is rising?

It is not easy to master our emotions while assessing information that matters to us, not least because our emotions can lead us astray in different directions. Gould realised he hadn't been thinking straight because of the initial shock – but how could he be sure, when he spotted those signs of hope in the statistics, that he wasn't now in a state of denial? He couldn't. With hindsight, he wasn't: he lived for another twenty years, and died of an unrelated condition.

We don't need to become emotionless processors of numerical information – just noticing our emotions and taking them into account may often be enough to improve our judgement. Rather than requiring superhuman control over our emotions, we need simply to develop good habits. Ask yourself: how does this information make me feel? Do I feel vindicated or smug? Anxious, angry or afraid? Am I in denial, scrambling to find a reason to dismiss the claim?

I've tried to get better at this myself. A few years ago, I shared a graph on social media which showed a rapid increase in support for same-sex marriage. As it happens, I have strong feelings about the matter and I wanted to share the good news. Pausing just long enough to note that the graph seemed to come from a reputable newspaper, I retweeted it.

The first reply was 'Tim – have you looked at the axes on that graph?' My heart sank. Five seconds looking at the graph would have told me that it was inaccurate, with the time scale a mess that distorted the rate of progress. Approval for marriage equality was increasing, as the graph showed, but I should have clipped it for my 'bad data visualisation' file rather than eagerly sharing it with the world. My emotions had got the better of me.

I still make that sort of mistake – but less often, I hope.

I've certainly become more cautious – and more aware of the behaviour when I see it in others. It was very much in evidence in the early days of the coronavirus epidemic, as helpful-seeming misinformation spread even faster than the virus itself. One viral post – circulating on Facebook and email newsgroups – all-too-confidently explained how to distinguish between Covid-19 and a cold, reassured people that the virus was destroyed by warm weather, and incorrectly advised that iced water was to be avoided, while warm water kills any virus. The post, sometimes attributed to 'my friend's uncle', sometimes to 'Stanford hospital board' or some blameless and uninvolved paediatrician, was occasionally accurate but generally speculative and misleading. Yet people – normally sensible people – shared it again and again and again. Why? Because they wanted to help others. They felt confused, they saw apparently useful advice, and they felt impelled to share. That impulse was only human, and it was well-meaning – but it was not wise.[8]

Before I repeat any statistical claim, I first try to take note of how it makes me feel. It's not a foolproof method against tricking myself, but it's a habit that does little harm and is sometimes a great deal of help. Our emotions are powerful. We can't make them vanish, and nor should we want to. But we can, and should, try to notice when they are clouding our judgement.

In 2011, Guy Mayraz, then a behavioural economist at the University of Oxford, conducted a test of wishful thinking.[9]

Mayraz showed his experimental subjects a graph of a price rising and falling over time. These graphs were actually historical snippets from the stock market, but Mayraz told people that the graphs showed recent fluctuations in the price of wheat. He asked each person to make a forecast of where the price would move next – and offered them a reward if their forecasts came true.

But Mayraz had also divided his experimental participants into two categories. Half of them were told that they were 'farmers', who would be paid extra if wheat prices were high. The rest were 'bakers', who would earn a bonus if wheat was cheap. So the subjects might earn two separate payments: one for making an accurate forecast, and the second a windfall if the price of wheat happened to move in their direction. Yet Mayraz found that the prospect of the windfall influenced the forecast itself. The farmers hoped that the price of wheat would rise, and they also *predicted* that the price of wheat would rise. The bakers hoped for – and predicted – the opposite. This is wishful thinking in its purest form: letting our reasoning be swayed by our hopes.

Another example was produced by economists Linda Babcock and George Loewenstein, who ran an experiment in which participants were given evidence from a real court case about a motorbike accident. They were then randomly assigned to play the role of plaintiff's attorney (arguing that the injured motorcyclist should receive $100,000 in damages) or defence attorney (arguing that the case should be dismissed or the damages should be low).

The experimental subjects were given a financial incentive to argue their side of the case persuasively and to reach an advantageous settlement with the other side. They were also given a separate financial incentive to accurately guess what damages the judge in the real case had actually awarded. Their predictions should have been unrelated to their role-playing, but again, their judgement was strongly influenced by what they hoped would be true.*[10]

Psychologists call this 'motivated reasoning'. Motivated reasoning is thinking through a topic with the aim, conscious or unconscious, of reaching a particular kind of conclusion. In a football game, we see the fouls committed by the other team but overlook the sins of our own side. We are more likely to notice what we want to notice.[11]

Perhaps the most striking example of this is among people who deny that the human immunodeficiency virus, HIV, causes AIDS. Some deny that HIV exists at all, but in any case HIV denialism implies rejecting the standard, and now highly effective, treatments. Some prominent believers in this idea have, tragically, doomed themselves and their children to death – but it must have been a comforting belief, particularly in the years when treatments for the condition were less effective and carried more severe side effects than they do today. One might assume that such a tragic belief would be vanishingly rare, but perhaps not. One survey of gay and bisexual men in the United States found that almost half believed HIV did not cause AIDS and more than half believed the standard treatments did more harm than good. Other surveys of people living with AIDS found the prevalence of denialist views at 15 to 20 per cent. These surveys weren't rigorous randomised samples, so I would not take the precise numbers too seriously. However, it's clear evidence that large numbers of people reject the scientific consensus in a way that could put them in real danger.[12]

I could see wishful thinking in operation in March

---

* In both cases it's conceivable that people were swayed less by the modest financial incentive and more by the emotional power of the role they were being asked to adopt. Either way, taking a particular perspective on the situation proved to be a strong influence on the decisions they made.

2020, too, when researchers at the University of Oxford published a 'tip of the iceberg' model of the pandemic. That model suggested that the coronavirus might be much more widespread but less dangerous than we thought, which had the joyful implication that the worst would soon be over. It was a minority view among epidemiologists, because the data detective work being done at that point saw little evidence that the vast majority of people had negligible symptoms. Indeed, one of the central points of the Oxford group was that we desperately needed better data to figure out the truth. That, however, was not the message that caught on. Instead, people widely shared the 'good news', because it was the kind of thing we all wanted to be true.[13]

Wishful thinking isn't the only form of motivated reasoning, but it is a common one. We believe in part because we want to. A person who is HIV-positive would find it comforting to believe that the virus does not lead to AIDS and cannot be passed to breastfeeding children. A 'farmer' wants to be accurate in his forecast of wheat prices, but he also wants to make money, so his forecasts are swayed by his avarice. A political activist wants the politicians she supports to be smart and witty and incorruptible. She'll go to some effort to ignore or dismiss evidence to the contrary.

And an art critic who loves Vermeer is motivated to conclude that the painting in front of him is not a forgery, but a masterpiece.

It was wishful thinking that undid Abraham Bredius. The art historian had a weak spot: his fascination with Vermeer's religious paintings. Only two existed. He had discovered one of them himself: *The Allegory of Faith*. He still owned it. The other, *Christ in the House of Martha and Mary*, was the only

Vermeer known to portray a scene from the Bible. Bredius had assessed it in 1901 and concluded quite firmly that it was not a Vermeer. Other critics disagreed, and eventually everyone reached the conclusion that Bredius had been wrong, including Bredius himself.

Stung by that experience, Bredius was determined not to repeat his mistake. He knew and loved Vermeer better than any man alive, and was always on the lookout for a chance to redeem himself by correctly identifying the next discovery of a Vermeer masterpiece.

And Bredius had become fascinated by the gap between the early, biblical *Martha and Mary* and Vermeer's more characteristic works, which were painted some years later. What lurked undiscovered in that gap? Wouldn't it be wonderful if another biblical work were found after all these years?

Bredius had another pet theory about Vermeer. The idea was that the Dutch master had, as a young man, travelled to Italy and been inspired by the religious works of the great Italian master Caravaggio. This was conjecture; not much was known about Vermeer's life. Nobody knew if he had ever seen a Caravaggio.

Van Meegeren knew all about Bredius's speculations. He painted *Emmaus* as a trap. It was a big, beautiful canvas, on a biblical theme, and – just as Bredius had argued all along – the composition was a homage to Caravaggio. Van Meegeren had planted some Vermeer-like touches in the painting, using seventeenth-century props. The bread that Christ is breaking is highlighted, just like that famous pearl earring, with thick dots of white paint called *pointillés*. And the paint was hard and cracked with age.

Bredius had no doubts. Why would he? Van Meegeren's stooge, Gerard Boon, wasn't just showing Bredius a painting: Boon was showing him evidence that he had been right all along. In the final years of his life, the old man had found the missing link at last. Bredius wanted to believe, and because he was an expert, he had no trouble in summoning up reasons to support his conclusion.

Those tell-tale *pointillés* on the bread, for instance: the white dots seem a bit clumsy to the untrained eye but they reminded Bredius of Vermeer's highlights on that tempting loaf of bread in *The Milkmaid*. The fact that the composition echoed Caravaggio would have been lost on a casual viewer, but leaped off the canvas under Bredius's gaze. He would have picked up other clues that *Emmaus* was the real thing. He would have noted the genuine seventeenth-century vase that van Meegeren had used as a prop. There were seventeenth-century pigments, too, or as close as possible. Van Meegeren had expertly duplicated Vermeer's colour palette. There was the canvas itself: an expert such as Bredius could spot a nineteenth- or twentieth-century forgery simply by looking at the back of the painting and noting that the canvas was too new. Van Meegeren knew this. He had painted his work on a seventeenth-century canvas, carefully scrubbed of its surface pigments but retaining the undercoat and its distinctive pattern of cracking.

And then there was the simplest test of all: was the paint soft? The challenge for anyone who wants to forge an old master is that oil paints take half a century to dry completely. If you dip a cotton bud into some pure alcohol and gently rub the surface of an oil painting, then the cotton may come away stained with pigments. If it does, the painting is a modern fake. Only after several decades will the paint harden enough to pass this test.

Bredius had identified fakes using this method before – but the paint on *Emmaus* stubbornly refused to yield its pigment. This gave Bredius an excellent reason to believe that *Emmaus* was old, and therefore genuine. Van Meegeren had fooled him with a brilliant piece of amateur chemistry, the result of many months of experimentation. The forger had figured out a way to mix seventeenth-century oil paints with a brand-new material: phenol formaldehyde, a resin that when heated at 105°C for two hours turned into one of the first plastics, Bakelite. No wonder the paint was hard and unyielding: it was infused with industrial plastic.

Bredius had half a dozen subtle reasons to believe that *Emmaus* was a Vermeer. They were enough to dismiss one glaring reason to believe otherwise: that the picture doesn't look like anything else Vermeer ever painted.

Take another look at that extraordinary statement from Abraham Bredius: 'We have here – I am inclined to say – the masterpiece of Johannes Vermeer of Delft . . . quite different from all his other paintings and yet every inch a Vermeer.'

'Quite different from all his other paintings' – shouldn't that be a warning? But the old man desperately wanted to believe that this painting was the Vermeer he'd been looking for all his life, the one that would provide the link back to Caravaggio himself. Van Meegeren set a trap into which only a true expert could stumble. Wishful thinking did the rest.

Abraham Bredius bears witness to the fact that experts are not immune to motivated reasoning. Under some circumstances their expertise can even become a disadvantage. The French satirist Molière once wrote, 'A learned fool is more foolish than an ignorant one.' Benjamin Franklin commented, 'So convenient a thing is it to be a reasonable creature, since it enables us to find or make a reason for everything one has a mind to.'

Modern social science agrees with Molière and Franklin:

people with deeper expertise are better equipped to spot deception, but if they fall into the trap of motivated reasoning, they are able to muster more reasons to believe whatever they really wish to believe.

One recent review of the evidence concluded that this tendency to evaluate evidence and test arguments in a way that's biased towards our own preconceptions is not only common, but just as common among intelligent people. Being smart or educated is no defence.[14] In some circumstances it may even be a weakness.

One illustration of this is a study published in 2006 by two political scientists, Charles Taber and Milton Lodge. Taber and Lodge were following in the footsteps of Kari Edwards and Edward Smith, whose work on politics and doubt we encountered in the introduction. As with Edwards and Smith, they wanted to examine the way Americans reasoned about controversial political issues. The two they chose were gun control and affirmative action.

Taber and Lodge asked their experimental participants to read a number of arguments on either side and to evaluate the strength and weakness of each argument. One might hope that being asked to review these pros and cons might give people more of a shared appreciation of opposing viewpoints; instead, the new information pulled people further apart. This was because people mined the information they were given for ways to support their existing beliefs. When invited to search for more information, people would seek out data that backed their preconceived ideas. When invited to assess the strength of an opposing argument, they would spend considerable time thinking up ways to shoot it down.

This isn't the only study to reach this sort of conclusion, but what's particularly intriguing about Taber and Lodge's experiment is that expertise made matters worse.* More sophisticated participants in the experiment found more material to back up their preconceptions. More surprisingly, they found less material that contradicted them – as though they were using their expertise actively to avoid uncomfortable information. They produced more arguments in favour of their own views, and picked up more flaws in the other side's arguments. They were vastly better equipped to reach the conclusion they had wanted to reach all along.[15]

Of all the emotional responses we might have, the most politically relevant are motivated by partisanship. People with a strong political affiliation want to be on the right side of things. We see a claim, and our response is immediately shaped by whether we believe 'that's what people like me think'.

Consider this claim about climate change: 'human activity is causing the Earth's climate to warm up, posing serious risks to our way of life'. Many of us have an emotional reaction to a claim like that; it's not like a claim about the distance to Mars. Believing it or denying it is part of our identity; it says something about who we are, who our friends are, and the sort of world we want to live in. If I put a claim about climate change in a news headline, or in a graph designed to be shared on social media, it will attract attention and engagement not because it is true or false but because of the way people feel about it.

If you doubt this, ponder the findings of a Gallup poll conducted in 2015. It found a huge gap between how much Democrats and Republicans in the United States worried about climate change. What rational reason could there be for that?

* Political expertise in this experiment was measured by asking people questions about the workings of US government – for example, how many congressional votes are needed to override a presidential veto?

Scientific evidence is scientific evidence. Our beliefs around climate change shouldn't skew left and right. But they do.[16]

This gap became wider the more education people had. Among those with no college education, 45 per cent of Democrats and 22 per cent of Republicans worried 'a great deal' about climate change. Yet among those with a college education, the figures were 50 per cent of Democrats and 8 per cent of Republicans. A similar pattern holds if you measure scientific literacy: more scientifically literate Republicans and Democrats are further apart than those who know very little about science.[17]

If emotion didn't come into it, surely more education and more information would help people to come to an agreement about what the truth is – or at least, the current best theory? But giving people more information seems actively to polarise them on the question of climate change. This fact alone tells us how important our emotions are. People are straining to reach the conclusion that fits with their other beliefs and values – and, like Abraham Bredius, the more they know, the more ammunition they have to reach the conclusion they hope to reach.

Psychologists call one of the processes driving this polarisation 'biased assimilation'. Imagine that you happen to encounter a magazine article that is discussing what we know about the effects of the death penalty. You're interested in the topic and so you read on, encountering the following brief account of a research study:

> Researchers Palmer and Crandall compared murder rates in 10 pairs of neighboring states with different capital punishment laws. In 8 of the 10 pairs, murder rates were higher in the state with capital punishment. This research opposes the deterrent effect of the death penalty.

What do you think? Does that seem plausible?

If you're opposed to the death penalty, then it probably does. But if you're in favour of the death penalty, doubts might start to creep in – those kind of doubts that we've already seen were so powerful in the case of tobacco. Was this research professionally conducted? Did they consider alternative explanations? How did they handle their data? In short, do Palmer and Crandall really know what they're doing, or are they a pair of hacks?

Palmer and Crandall won't be offended by your doubts. The duo do not exist. They were dreamed up by three psychologists, Charles Lord, Lee Ross and Mark Lepper. In 1979, Lord, Ross and Lepper conducted an experiment that was designed to explore how people thought through arguments they felt passionately about. The researchers rounded up experimental subjects with strong views in favour of, or against, the death penalty. They showed the experimental subjects summaries of two imaginary studies. One of these made-up studies demonstrated that the death penalty deterred serious crime; the other, by the fictitious researchers Palmer and Crandall, showed the opposite.[18]

As one might expect, the experimental subjects were inclined to dismiss studies that contradicted their cherished beliefs. But Lord and his colleagues discovered something more surprising: the more detail people were presented with – graphs, research methods, commentary by other fictional academics – the easier they found it to disbelieve unwelcome evidence. If doubt is the weapon, detail is the ammunition.

When we encounter evidence that we dislike, we ask ourselves, '*Must* I believe this?' More detail will often give us more opportunity to find holes in the argument. And when we encounter evidence that we approve of, we ask a different question: '*Can* I believe this?' More detail means more toeholds on to which that belief can cling.[19]

The counterintuitive result is that presenting people with a detailed and balanced account of both sides of the argument may actually push people away from the centre rather than pull them in. If we already have strong opinions, then we'll seize upon welcome evidence, but we'll find opposing data or arguments irritating. This 'biased assimilation' of new evidence means that the more we know, the more partisan we're able to be on a fraught issue.

Maybe this sounds absurd. Don't we all want to figure out the truth? We certainly should when it will affect us personally – and the tragic case of HIV/AIDS denialism indicates that some people will go to extraordinary lengths to reject ideas that are uncomfortable and unwelcome, even if those ideas could save their lives. Wishful thinking can be astonishingly powerful.

But often being right doesn't have such profound consequences. On many questions, reaching a factually incorrect conclusion causes us no harm at all. It can even help us.

To see why, ponder an issue where most people would agree that there is no objective 'truth' at all: the moral difference between eating beef, eating pork and eating dog. Which of these practices you think is right and which is wrong depends mostly on your culture. Few people will care to discuss the underlying logic of the matter. It's better to fit in.

Less obviously, the same is often true of arguments where there *is* a correct answer. In the case of climate change, there is an objective truth even if we are unable to discern it with perfect certainty. But as you are one individual among nearly 8 billion on the planet, the environmental consequences of what you happen to think are irrelevant. With a handful of exceptions – say, if you're the president of China – climate change is going to take its course regardless of what you say or do. From a self-centred point of view, the practical cost of being wrong is close to zero.

The social consequences of your beliefs, however, are real and immediate.

Imagine that you're a barley farmer in Montana, and hot, dry summers are ruining your crop with increasing frequency. Climate change matters to you. And yet rural Montana is a conservative place, and the words 'climate change' are politically charged. Anyway, what can you personally do about it? Here's how one farmer, Eric Somerfeld, threads that needle:

> In the field, looking at his withering crop, Somerfeld was unequivocal about the cause of his damaged crop – 'climate change.' But back at the bar, with his friends, his language changed. He dropped those taboo words in favor of 'erratic weather' and 'drier, hotter summers' – a not-uncommon conversational tactic in farm country these days.[20]

If Somerfeld lived in Portland, Oregon, or Brighton, England, he wouldn't need to be so circumspect at his local tavern – he'd be likely to have friends who took climate change very seriously indeed. But then those friends would quickly ostracise someone else in the social group who went around loudly claiming that climate change is a Chinese hoax.

So perhaps it is not so surprising after all to find educated Americans poles apart on the topic of climate change. Hundreds of thousands of years of human evolution have wired us to care deeply about fitting in with those around us. This helps to explain the findings of Taber and Lodge that better-informed people are actually more at risk of motivated reasoning on politically partisan topics: the more persuasively we can make the case for what our friends already believe, the more our friends will respect us.

HIV denialism shows we're capable of being tragically wrong even in matters of life and death. But it's far easier to lead ourselves astray when the practical consequences of being wrong are small or non-existent, while the social consequences of being 'wrong' are severe. It's no coincidence that this describes many controversies that divide along partisan lines.

It's tempting to assume that motivated reasoning is just something that happens to other people. I have political principles; you're politically biased; he's a fringe conspiracy theorist. But we'd be wiser to acknowledge that we all think with our hearts rather than our heads sometimes.

Kris De Meyer, a neuroscientist at King's College, London, shows his students a message describing an environmental activist's problem with climate change denialism:

> To summarize the climate deniers' activities I think we can say that:
>
> (1) Their efforts have been aggressive while ours have been defensive.
> (2) The deniers' activities are rather orderly – almost as if they had a plan working for them.
>
> I think the denialist forces can be characterized as dedicated opportunists. They are quick to act and seem to be totally unprincipled in the type of information they use to attack the scientific community. There is no question, though, that we have been inept in getting our side of the story, good though it may be, across to the news media and the public.[21]

The students, all committed believers in climate change, outraged at the smokescreen laid down by the cynical and anti-scientific deniers, nod in recognition. Then De Meyer reveals the source of the text. It's not a recent email. It's taken, almost word for word, from an infamous internal memo written by a cigarette marketing executive in 1968. The memo is complaining not about 'climate deniers' but about 'anti-cigarette forces', but otherwise no changes were required. You can use the same language, the same arguments, and perhaps even have the same conviction that you're right, whether you're arguing (rightly) that climate change is real or (wrongly) that the cigarette–cancer link is not.

(Here's an example of this tendency that, for personal reasons, I can't help but be sensitive about. My left-leaning, environmentally conscious friends are justifiably critical of ad hominem attacks on climate scientists. You know the kind of thing: claims that scientists are inventing data because of their political biases or because they're scrambling for funding from big government. In short, smearing the person rather than engaging with the evidence. Yet the same friends are happy to embrace and amplify the same kind of tactics when they're used to attack my fellow economists: that we're inventing data because of our political biases, or scrambling for funding from big business. I tried to point out the parallel to one thoughtful person, and got nowhere. She was completely unable to comprehend what I was talking about. I'd call this a 'double standard', but that would be unfair – it would suggest that it was deliberate. It's not. It's an unconscious bias that's easy to see in others and very hard to see in ourselves.)*

Our emotional reaction to a statistical or scientific claim isn't a side issue. Our emotions can, and often do, shape our

* I'm quite sure that I'm guilty, too. I just can't see exactly how.

beliefs more than any logic. We are capable of persuading ourselves to believe strange things, and to doubt solid evidence, in service of our political partisanship, our desire to keep drinking coffee, our unwillingness to face up to the reality of our HIV diagnosis, or any other cause that invokes an emotional response.

But we shouldn't despair. We can learn to control our emotions – that is part of the process of growing up. The first simple step is to notice those emotions. When you see a statistical claim, pay attention to your own reaction. If you feel outrage, triumph, denial, pause for a moment. Then reflect. You don't need to be an emotionless robot, but you could and should think as well as feel.

Most of us do not actively wish to delude ourselves, even when that might be socially advantageous. We have motives to reach certain conclusions, but facts matter too. Lots of people would like to be movie stars, billionaires, or immune to hangovers, but very few people believe that they actually are. Wishful thinking has limits. The more we get into the habit of counting to three and noticing our knee-jerk reactions, the closer to the truth we are likely to get.

For example, one survey, conducted by a team of academics, found that most people were perfectly able to distinguish serious journalism from fake news, and also agreed that it was important to amplify the truth, not lies. Yet the same people would happily share headlines such as 'Over 500 "Migrant Caravaners" Arrested With Suicide Vests', because at the moment at which they clicked 'share', they weren't stopping to think. They weren't thinking, 'is this true?' and they weren't thinking, 'do I think the truth is important?'. Instead, as they skimmed the internet in that state of constant distraction that we all recognise, they were carried away with their emotions and their partisanship. The good news is that simply pausing for a moment to reflect was all it took to filter out a lot of the misinformation. It doesn't take much; we can all do it. All we need to do is acquire the habit of stopping to think.[22]

Another study found that people who were best able to distinguish real from fake news were also the people who scored highly on what is called a 'cognitive reflection test'.[23] These tests – created by Shane Frederick, a behavioural economist, and made famous by Daniel Kahneman's book *Thinking, Fast and Slow* – ask questions such as:

> A bat and ball cost $1.10, and the bat costs a dollar more than the ball. How much does the ball cost?

> and

> A lake contains a patch of lily pads which doubles in size each day. If it takes 48 days for the patch to cover the entire lake, how long would it take for the patch to cover half of the lake?*

Many people get the answers to these questions wrong the first time they hear them, but what's required to reach the correct solution isn't intelligence or mathematical training, but pausing for a moment to double-check your gut reaction. Shane Frederick points out that noticing your initial error is usually all that's necessary to solve the problem.[24]

The cognitive reflection questions invite us to leap to the wrong conclusion without thinking. But so, too, do

---

* The answers: five cents, and forty-seven days.

Perhaps the second question is less of a stumbling block than once it was. The lily patch is growing exponentially, and we have all received a hard lesson from the coronavirus in what exponential growth looks like.

inflammatory memes or tub-thumping speeches. That's why we need to be calm. And that is also why so much persuasion is designed to arouse us – our lust, our desire, our sympathy or our anger. When was the last time Donald Trump, or for that matter Greenpeace, tweeted something designed to make you pause in calm reflection? Today's persuaders don't want you to stop and think. They want you to hurry up and feel.

Don't be rushed.

Han van Meegeren had been arrested almost immediately after German occupation ended. He should have been prosecuted and punished for collaboration with the Nazis.

The wily forger had prospered mightily under Nazi occupation. He owned several mansions. While Amsterdam starved during the war, he hosted regular orgies at which prostitutes helped themselves to fistfuls of jewels. If he wasn't actually a Nazi himself, he went to extraordinary lengths to behave like one. He was friends with Nazis, and he bent over backwards to celebrate Nazi ideology.

Van Meegeren illustrated and published a lavishly evil book called *Teekeningen 1*, full of grotesque anti-Semitic poetry and illustrations, using the Nazi iconography and colours. He spared no expense in the printing of the book, and no wonder, given whom he imagined might read it. A copy was hand-delivered to Adolf Hitler, with a handwritten dedication in artist's charcoal: 'To My beloved Führer in grateful tribute – Han van Meegeren'.

It was found in Hitler's library.

To understand what happened next, we need to understand emotion rather than logic. The Dutch were disillusioned with themselves after five years of German occupation. Anne Frank was just the most famous of the huge number of Jews to have been deported from the Netherlands and murdered, but it is less well known that a far higher proportion of Dutch Jews were deported than those living in France or Belgium.[25] Van Meegeren, of course, was yet another collaborator. But in the wake of the war, the Dutch had become tired of parading such men through their courts, month after month. They desperately wanted a more inspiring story – just as Abraham Bredius desperately wanted to find a Caravaggio-esque Vermeer. Yet again, van Meegeren produced what was wanted: this time, a light-hearted tale of boldness and trickery in which a Dutchman had struck back against the Nazis.

The men responsible for prosecuting van Meegeren soon became his unwitting accomplices. They arranged an absurd publicity stunt where he 'proved' that he was a forger rather than a traitor by painting a picture in the style of *Emmaus*. One breathless headline reported, 'He Paints for His Life'. Newspapers in the Netherlands and around the world couldn't tear their gaze away from the great showman.

Then came the trial, a media circus in which the charismatic van Meegeren was the ringmaster. He spun his story: that he had only forged the art to prove his worth as an artist, and to unmask the art experts as fools. When the judge reminded him that he had sold the fakes for high prices, he replied, 'Had I sold them for low prices, it would have been obvious they were fake.' The courtroom laughed; van Meegeren had them all spellbound. A man who should have been viewed as a traitor reshaped his reputation into that of a patriot, even a hero. He manipulated the emotions of the Dutch people, as he had manipulated the emotions of Abraham Bredius before the war.

It wasn't just the Dutch who swallowed the story of the man who played Göring for a fool. Van Meegeren found plenty of people who were delighted to play up the deliciousness of the story. Early biographers of van Meegeren made him out to be a misunderstood trickster, hurt by the unjust rejections of his own art, but happy to outsmart his country's occupiers. One oft-reported story is that Göring, awaiting trial in Nuremberg, when told that he had been duped by van Meegeren, 'looked as if for the first time he had discovered there was evil in the world'. When you hear that anecdote it's almost impossible to resist repeating it. But like the *pointillés* on the bread, it's a telling detail that is just as false.

If only Hitler's personally inscribed copy of *Teekeningen 1* had been discovered before van Meegeren's trial, the story of the daring little forger would have dissolved. The truth about van Meegeren would have been obvious. Or would it?

The discomfiting truth about *Teekeningen 1* is that the dedicated copy in Hitler's library had been found almost immediately. *De Waarheid*, a Dutch resistance newspaper, had announced the discovery on 11 July 1945. It just didn't matter; nobody wanted to know. Van Meegeren waved the truth away, claiming that he had signed hundreds of copies of the book and the dedication must have been added by someone else. In a modern setting he might have dismissed the newspaper report as 'fake news'.

It was a ludicrous excuse, but van Meegeren had managed to hypnotise his prosecutors just as he had hypnotised Bredius, by distracting them with interesting details and selling them a story they wanted to believe.

In his closing statement to the court he claimed again that he hadn't done it for the money, which had brought him nothing but trouble. It was a bold claim: we should remember that while wartime Amsterdam went hungry, van Meegeren liked to accessorise his mansions with prostitutes, jewels, and prostitutes draped with jewels. No matter: the newspapers and the public lapped up his story.

After being found guilty of forgery, van Meegeren was

cheered as he left the courtroom. He had pulled off an even more audacious con – a fascist and a fraud successfully presented himself as a cheeky hero of the Dutch people. Abraham Bredius desperately wanted a Vermeer. The Dutch public desperately wanted symbols of resistance to the Nazis. Han van Meegeren knew how to give people what they wanted.

Before serving a day of his sentence, van Meegeren died, on 30 December 1947, of a heart attack. An opinion poll conducted a few weeks earlier had found him to be (except for the Prime Minister) the most popular man in the country.

If wishful thinking can turn a rotten fake into a Vermeer, or a sleazy Nazi into a national hero, then it can turn a dubious statistic into solid evidence, and solid evidence into fake news. But it doesn't have to. There is hope. We're about to go on a journey of discovery, finding out how numbers can make the world add up. The first step, then, is to stop and think when we are being presented with a new piece of information, to examine our emotions and to notice if we're straining to reach a particular conclusion.

When we encounter a statistical claim about the world, and are thinking of sharing it on social media or typing a furious rebuttal, we should instead ask ourselves: 'How does this make me feel?'*

We should do this not just for our own sake, but as a social duty. We've seen how powerful social pressure can be in influencing what we believe and how we think. When we slow down, control our emotions and our desire to signal partisan affiliation, and commit ourselves to calmly weighing the facts, we're not just thinking more clearly – we are

---

* A follow-up question might also be worth asking: *why* does it make me feel that way?

also modelling clear thinking for others. It is possible to take a stand not as a member of a political tribe but as someone who is willing to reflect and reason in a fair-minded manner. I want to set that sort of example. I hope that you do, too.

Van Meegeren understood all too well that how we feel shapes what we think. Yes, expertise and technical knowledge matter, but the technical side of dealing with numbers will come in the chapters that follow. If we don't master our emotions, whether they are telling us to doubt or telling us to believe, we're in danger of fooling ourselves.

RULE TWO

Ponder your personal experience

> In a bird's eye view you tend to survey everything ... In a worm's eye view you don't have that advantage of looking at everything. You just see whatever is close to you.
>
> —MUHAMMAD YUNUS[1]

As I settled into presenting *More or Less*, I felt I had a dream gig. Debunking numerical nonsense in the news was fun, and by looking through the statistical telescope I was constantly seeing new and interesting things. There was, however, a snag: every time I travelled to the BBC studios to record the programme, I felt that my personal experience was contradicting some credible-seeming statistics.

Let me explain. The commute wasn't the world's most glamorous journey. To get to White City in west London from Hackney in east London, I'd scurry across a busy road, hop on to a busy double-decker bus, and watch the traffic as we moved slowly towards Bethnal Green, the underground station. If the bus had been busy, the tube train was busier. It made a can of sardines look roomy. I'd join a crowd of hopeful passengers on the platform, waiting for a Central Line train to arrive with enough space to squeeze on. That was by no means a certainty. We'd often have to wait for the second or third train before being able to wriggle between the less-than-delighted passengers who'd ridden in from further east. Getting a seat was out of the question.

It was this experience that challenged my view that numbers make the world add up, because when I looked at the statistics about how busy London's public transport actually was, they flatly contradicted the evidence of my own eyes – and on warmer, sweatier days, my own nose. Those statistics showed that the average occupancy of a London bus was around twelve people, a tiny number compared with the sixty-two seats available on the double-decker bus I rode every morning.[2] That felt completely wrong. Some days I felt there were more than twelve people within arm's reach, let alone on the bus.

The tube occupancy rates made even less sense. According to Transport for London, the 'crush capacity' of one of those tube trains is more than a thousand people.[3] But the average occupancy? Less than 130.[4] What? You could *lose* 130 people on a Central Line tube train. You could squeeze them on to a single carriage and leave the other seven completely empty. And that's not the occupancy at quiet moments – it's the average. Was I really supposed to believe that these statistics – twelve people on a bus, 130 people on a train – reflected reality? Surely not, not when every single time I took a trip to work I could not only barely get on to the train, I would sometimes struggle to get on to the platform. The trains *must* be busier than the statistics showed.

In the studio, I was singing the praises of statistical thinking. But on the way to the studio, my everyday experience told me that these particular statistics must be wrong.

The contradiction between what we see with our own eyes and what the statistics claim can be very real. In the previous chapter we discovered that it is important not to be fooled by our personal feelings. As I'm a self-confessed data detective, you might expect me to say the same about our personal experiences, too. After all, who are you going to believe? A trusty spreadsheet, or your own lying eyes?

The truth is more complicated. Our personal experiences should not be dismissed along with our feelings, at least not without further thought. Sometimes the statistics give us a vastly better way to understand the world; sometimes they mislead us. We need to be wise enough to figure out when the statistics are in conflict with everyday experience – and in those cases, which to believe.

So what should we do when the numbers tell one story, and day-to-day life tells us something different? That's what this chapter is about.

We might start by being curious about where the statistics come from. In the case of my commute, the numbers are published by Transport for London (TfL), the government organisation which oversees London's roads and public transport. But how do the fine folk of TfL know for sure how many people are on a bus or a tube train? It's a good question, and the answer is: they don't. They can, however, make a good guess. Years ago, estimates were based on paper surveys, carried out by researchers standing at bus stops or in tube stations with a clipboard, or handing out questionnaires. Clearly this was a ponderous method, although it is unlikely that it introduced enough errors to explain the huge disparity between my experience and the official occupancy figures.

In any case, in the age of contactless payments it's much

easier to estimate passenger numbers. The vast majority of bus journeys are made by people tapping an identifiable contactless chip on a bank card, a TfL Oyster card or a smartphone. The data scientists at TfL can see where and when these devices are being used. They still have to make an educated guess as to when you get off the bus, but this is often possible – for example, they might see you make the return journey from the same area later. Or they might see that you had used your card on a connecting service: whenever I tapped into the tube network at Bethnal Green, one minute after the bus I'd been riding on arrived in the area, TfL could conclude with confidence that I'd been on the bus until the stop at Bethnal Green, but no further.

On the London Underground, people tap in and out, but TfL still does not know what route commuters took across the network, which often offers several plausible alternatives. TfL thus still doesn't know how busy particular trains are. Again, they can make an educated guess, using occasional paper-based surveys to supplement their judgement as to how passengers are choosing to get around.

The estimates will soon be more accurate yet. On 8 July 2019, TfL switched on a system to use wi-fi networks to measure how crowded are different parts of the London Underground. The more phones are trying to connect to wi-fi, the busier the pinch point in a particular station. This system promises to let TfL spot overcrowding and other problems in real time. (I spoke to the data team at TfL the day after this system was switched on. They were adorably excited.)[5]

The statistics, then, are at least plausible. We can't simply dismiss them as mistaken.

The next step is to look for reasons why our personal experience might be so different. In the case of my commute, the obvious starting point is that I was travelling at a busy time of day, on one of the busiest sections of the tube network. No wonder it was crowded.

But this particular rabbit-hole goes a little deeper. It's perfectly possible that most trains aren't crowded, and yet most people travel on crowded trains. For an extreme illustration, imagine a hypothetical train line with ten trains a day. One rush-hour train has a thousand people crammed on to it. All the other trains carry no passengers at all. What's the average occupancy of these trains? A hundred people – not far off the true figure on the London Underground. But what is the experience of the typical passenger in this scenario? Every single person rode on a crowded train.

The real situation on the London Underground isn't as extreme. There aren't many completely empty trains, but trains do sometimes run with very few passengers on them, particularly when they're running counter to the flow of commuters. Whenever they do, very few passengers will be around to witness it. Those statistics are telling the truth – but not the whole truth.

Of course, there are alternative ways to gauge the problem of overcrowding. Rather than measure the occupancy of the average train, you could measure the situation faced by the average passenger: out of a hundred passenger journeys, how many are on overcrowded trains? That would be a better way to measure the passenger experience – and indeed TfL are now refocusing their data collection and reporting to produce statistics that reflect the situation not of the trains, but of the passengers.

Yet there's no single objective measure of how busy the public transport network is. As a passenger, it seems to me that all the buses I'm on are well used. But TfL's statistics show, truthfully, that many buses are driving around largely empty. This is because buses don't just appear in the busiest areas by magic; when they reach the end of the route they have to turn round and go back again. TfL care about the low average occupancy of buses because those buses cost money, take up space on the roads, and emit pollution. The average occupancy is therefore an important metric for them.

In short, my own eyes told me something important and true about London's transport network. But the statistics told me something else, something equally important and equally true – and something I couldn't have known in any other way. Sometimes personal experience tells us one thing, the statistics tell us something quite different, and both are true.

That's not always the case, of course. Think back to the discovery that heavy cigarette smoking increased the risk of lung cancer by a factor of sixteen. Many people would have found reason from their personal experience to be sceptical of this finding. Perhaps your chain-smoking nonagenarian grandma is as fit as a fiddle, whereas the only person you know who died from lung cancer is your next-door neighbour's uncle and he never smoked a cigarette in his life.

On the face of it, this seems no different to the experience of my daily commute appearing to contradict TfL's statistics. But on closer inspection, in this case we do find reason to discard our personal experience and trust the statistical view. Though a factor of sixteen is hardly a small effect, lung cancer is itself scarce enough to confuse our intuitions. The world is full of patterns that are too subtle or too rare to detect by eyeballing them, and a pattern doesn't need to be very subtle or rare to be hard to spot without a statistical lens.

This is true of many medical conditions and treatments. When we feel bad – anything from a headache to depression, a sore knee to an unsightly spot – we seek solutions. My wife recently suffered from a sharp pain in her shoulder whenever she raised her arm; it was bad enough to make it hard for her to get dressed or reach something on a high shelf. After a while, she went to a physiotherapist, who diagnosed the problem and prescribed some uncomfortable exercises, which my wife diligently performed every day. After a few weeks, she told me, 'I think my shoulder is getting better.'

'Wow – looks like the physiotherapy worked!' I said.

'Maybe,' said my wife, who can spot me setting a statistical trap a mile off. 'Or maybe it would have got better by itself anyway.'

Indeed. From my wife's point of view, it didn't really matter. What she wanted was for her shoulder to heal, and the evidence of her own senses was the only relevant yardstick. But for the question of whether the exercises had caused the recovery, her personal experience wasn't much use – and from the point of view not of my wife but of future shoulder-pain sufferers, it's the question of causation that matters. We need to know whether those exercises tend to help, or whether there might be a better approach.

The same is true of any other treatment for any other problem, whether it's diet, therapy, exercise, antibiotics or painkillers: it's nice if we feel better, but future generations need to know whether we feel better because of the steps we've taken, or whether they were empty rituals that did no good, cost money, wasted time and produced unwelcome side effects. For this reason, we rely on randomised trials of any treatment, ideally compared against the best available treatment, or against a fake treatment called a placebo. It's not that our personal experience is irrelevant, it's that it can't give us the information we need to help those who come after us.

When personal experience and statistics seem to be in

conflict, a closer look at the situation may reveal particular reasons why personal experience is likely to be an unreliable guide. Consider the idea that the vaccination against measles, mumps and rubella (MMR) increases a child's risk of autism. It doesn't, but fewer than half of us are convinced of that.[6]

We can say with confidence that there is no such link thanks to the statistical perspective. Since autism is not common, we need to compare the experiences of many thousands of children who have received the vaccination, and those who have not. One major study, in Denmark, did exactly that. It followed 650,000 children. Most of them received an MMR vaccine at the age of fifteen months, and a follow-up at four years, but about 30,000 did not. About 1 per cent of children were then diagnosed with autism, and that was true both of the vaccinated and the unvaccinated children. (The unvaccinated children, of course, were at higher risk of contracting these dangerous diseases.)[7]

So why do many people remain sceptical? Part of the answer is a sad history of reckless publishing around the issue. But in part the doubts persist because many people have heard of children whose autism was diagnosed soon after an MMR vaccination, and whose parents think the MMR was to blame. Imagine taking your child for the vaccination, and soon afterwards receiving a diagnosis of autism. Would you connect the two? It would be hard not to wonder.

In fact, the prevalence of such anecdotes is not surprising because autism tends to be diagnosed at one of two ages: early signs of the condition are observable by paediatric nurses at around the age of fifteen months; if not picked up then, diagnosis often follows a child starting school.[8] And the two doses of the MMR vaccine are routinely given close to these ages. When we find a convincing explanation for why our personal experience sits uneasily with the statistical view, it should reassure us to set aside our doubts and trust the numbers.

A less fraught example is our relationship with television and other media. Many people on television are richer than you and me. Almost by definition, they are more famous than you and me. It's very likely that they are better-looking than you and me; they are certainly better-looking than me (I am on the radio for a reason). When we reflect on how attractive, famous and rich the typical person is, we can't help but have our assessment skewed by the fact that many of the people we know, we know through the media; they are attractive, famous and rich. Even if, on reflection, we realise that TV personalities aren't a random sample of the global population, it's hard to shake the feeling that they are.

Psychologists have a name for our tendency to confuse our own perspective with something more universal: it's called 'naive realism', the sense that we are seeing reality as it truly is, without filters or errors.[9] Naive realism can lead us badly astray when we confuse our personal perspective on the world with some universal truth. We are surprised when an election goes against us: everyone in our social circle agreed with us, so why did the nation vote otherwise? Opinion polls don't always get it right, but I can assure you they have a better track record of predicting elections than simply talking to your friends.

Naive realism is a powerful illusion. Consider the findings of a survey from the opinion pollster Ipsos MORI. MORI asked nearly 30,000 people across thirty-eight countries about a range of social issues, finding them – and, presumably, most of the rest of us – badly out of step with what credible statistics showed:[10]

(a) We're wrong about the murder rate. We think it's been rising since the year 2000. In most of the countries surveyed, it's been falling.
(b) We think deaths from terrorism have been higher in the past fifteen years than in the fifteen years before; they're down.
(c) We think that 28 per cent of prisoners are immigrants. Ipsos MORI reckons the true rate across the countries surveyed was 15 per cent.
(d) We think that 20 per cent of teenage girls give birth each year. This number strains biological credibility when you think about it. An eighteen-year-old has been a teenager for six years, so if each year she has a 20 per cent chance of having a baby, most eighteen-year-olds are mothers. (Those who aren't are balanced by the eighteen-year-olds who are mothers several times over.) Look around; is that really true? The correct figure, says Ipsos MORI, is that 2 per cent of teenage girls give birth each year.*
(e) We think that 34 per cent of people have diabetes; the true figure is 8 per cent.
(f) We think that 75 per cent of people have a Facebook account. The correct figure at the time of asking, 2017, was 46 per cent.

Why are our perceptions of the world so mistaken? It's hard to be sure, but a plausible first guess is that we're getting

* This is a reminder of how useful it is to stop and think. There is no advanced mathematics required to realise that the 20 per cent figure simply cannot be squared with our everyday experience. In some countries, people say they believe that 50 per cent of teenage girls give birth each year, which would imply young women typically enter adulthood with three children of their own.

our impressions from the media. It's not that a reputable newspaper or TV channel would actually give us the wrong data – although it has been known. The problem is that the news carries tales of lottery wins and fairy-tale romances, terrorist atrocities or gruesome assaults by strangers, and of course the latest trends, which are often not nearly as popular as they seem. None of these stories reflects everyday life; all of them are viscerally memorable and seem to take place in our living rooms. We form our impressions accordingly.

As the great psychologist Daniel Kahneman explained in *Thinking, Fast and Slow*: 'When faced with a difficult question, we often answer an easier one instead, usually without noticing the substitution.' Rather than asking 'Are terrorists likely to kill me?' we ask ourselves, 'Have I recently seen a news report about terrorism?' Instead of saying, 'Out of all the teenage girls I know, how many are already mothers?' we say, 'Can I think of a recent example of a news story about teenage pregnancy?'

These news reports are data, in a way. They're just not representative data. But they certainly influence our views of the world. To adapt Kahneman's terminology, they're 'fast statistics' – immediate, intuitive, visceral and powerful. 'Slow statistics', those based on a thoughtful gathering of unbiased information, aren't the ones that tend to leap into our minds. But as we shall see, there are ways to consume more of the slow stuff and have a more balanced diet of information as a result.

So far we've seen cases in which the ponderous-and-careful slow statistics are more trustworthy than the quick-and-dirty fast statistics, and situations in which both give us a useful angle on the world. But are there also cases where we should trust our personal impressions more than the data?

Yes. There are certain things that we cannot learn from a spreadsheet.

Consider Jerry Z. Muller's book, *The Tyranny of Metrics*. It's 220 pages long. The average chapter is 10.18 pages long and contains 17.76 endnotes. There are four cover endorsements and the book weighs 421 grams. But of course none of these numbers tells us what we want to know – which is what does the book say, and should we take it seriously? To understand the book you will need to read it, or trust the opinion of someone who has.

Jerry Muller takes aim at the problem with a certain kind of 'slow statistics' – those used as management metrics or performance targets. Statistical metrics can show us facts and trends that would be impossible to see in any other way, but often they're used as a substitute for relevant experience, by managers or politicians without specific expertise or a close-up view. For example, if a group of doctors collect and analyse data on clinical outcomes, they are likely to learn something together that helps them to do their jobs. But if the doctors' bosses then decide to tie bonuses or professional advancement to improving these numbers, unintended consequences will predictably occur. For example, several studies have found evidence of cardiac surgeons refusing to operate on the sickest patients for fear of lowering their reported success rates.[11]

In my book *Messy*, I spent a chapter discussing similar examples. There was the time the UK government collected data on how many days people had to wait for an appointment when they called their doctor, which is a useful thing to know. But then the government set a target to reduce the average waiting time. Doctors logically responded by refusing to take any advance bookings at all; patients had to phone up every morning and hope they happened to be among the first to get through. Waiting times became, by definition, always less than a day.

What happened when a widely consulted ranking of US colleges, the *US News and World Report*, rewarded more selective institutions? Over-subscribed universities scrambled to attract fresh applicants that they could reject, and thereby appear to be more selective.

Then there is the notorious obsession with the 'body count' metric, which was embraced by US Defense Secretary Robert McNamara during the Vietnam War. The more of the enemy you kill, reasoned McNamara, the closer you are to winning. This was always a dubious idea, but the body count quickly became an informal metric for ranking units and handing out promotions, and was therefore often exaggerated. And since it was sometimes easier to count enemies who were already dead than to kill anyone new, counting bodies became a military objective in itself. It was risky, and it was useless, but it responded to the skewed incentive McNamara had set.

This episode shows that statistics aren't always worth gathering – but you can appreciate why McNamara wanted them. He was trying to understand and control a distant situation, one he had no experience of as a soldier. A few years ago I interviewed General H. R. McMaster, an expert on the mistakes made in Vietnam. He told me that the army used to believe that 'situational understanding could be delivered on a computer screen'.

It could not. Sometimes you have to be there to understand – especially when a situation is fast-moving or contains soft, hard-to-quantify details, as is typically the case on the battlefield. The Nobel laureate economist Friedrich Hayek had a phrase for the kind of awareness it's hard to capture in metrics and maps: the 'knowledge of the particular circumstances of time and place'.

Social scientists have long understood that statistical metrics are at their most pernicious when they are being used to control the world, rather than try to understand it. Economists tend to cite their colleague Charles Goodhart, who wrote in 1975: 'Any observed statistical regularity will tend to collapse once pressure is placed upon it for control purposes.'[12] (Or, more pithily: 'When a measure becomes a target, it ceases to be a good measure.') Psychologists turn to Donald T. Campbell, who around the same time explained: 'The more any quantitative social indicator is used for social decision-making, the more subject it will be to corruption pressures and the more apt it will be to distort and corrupt the social processes it is intended to monitor.'[13]

Goodhart and Campbell were on to the same basic problem: a statistical metric may be a pretty decent proxy for something that really matters, but it is almost always a proxy rather than the real thing. Once you start using that proxy as a target to be improved, or a metric to control others at a distance, it will be distorted, faked or undermined. The value of the measure will evaporate.

In 2018, I visited China with my family. The trip taught me that I shouldn't need to favour either fast or slow statistics; the deepest understanding comes from melding them together.

The slow statistics tell a familiar story – familiar, at least, to economics geeks like me. Real income per person in China has increased ten-fold since 1990. Since the early 1980s, the number of extremely poor people there has fallen by more than three quarters of a billion – well over half the entire population of the country. China consumed more cement in a recent three-year period than the United States used in the entire twentieth century. On paper, it is the most dramatic explosion of economic activity in human history.

Yet seeing it with your own eyes is another experience entirely. Nothing in the statistics truly prepared me for a journey across Guangdong, the southern province of China that has been at the forefront of this growth. We started at Hong Kong – the ultimate high-rise city – and walked into its mainland twin, Shenzhen. Then in the shadow of the Ping An skyscraper, which dwarfs the Empire State Building, we caught a bullet train across the province.

Where London's tower blocks often stand alone or in groups of two or three, Shenzhen will have a cluster of a dozen identical monoliths, crammed with apartments, shoulder to shoulder. Next to that cluster, another dozen of a different design. Then another, and another. Here and there, in the distance across the haze, would be a Manhattan-esque cluster of larger skyscrapers. The towers marched on and on, all the way (or so it seemed to me) to the city of Guangzhou – forty-five minutes or so of high-speed travel through an infinite vista of concrete.

We ended the day much deeper into China, in the picture-postcard landscape of Yangshuo. But despite the idyllic surroundings, I couldn't sleep. The endless tower blocks scrolled through my mind. What if we had lost our six-year-old son in the middle of Guangdong? And my sleepless anxieties flitted back and forth between my family and the world. So many people. So much concrete. How could the planet possibly survive this?

Of course, there was nothing in this experience to contradict the economic data; the two perspectives on China's growth were perfectly complementary. But they felt very different. The 'slow statistics' required me to reflect and calculate, taking some effort to process the numbers and follow the logic of what they implied for modern China. The train journey delivered 'fast statistics' instead. It tapped into a different and more intuitive way of thinking, as I swiftly and automatically formed my impressions, compared Guangzhou to the cities I knew back home, and anxiously sensed the danger to those I love.*

Both ways of understanding the world have their own advantages, and their own traps. Muhammad Yunus, an economist, microfinance pioneer and winner of the Nobel Peace Prize, has contrasted the 'worm's eye view' of personal experience with the 'bird's eye view' that statistics can provide. The worm and the bird see the world very differently, and Professor Yunus is right to emphasise the advantage of seeing it up close.

But birds see a lot, too. Professor Yunus, paying close attention to the lives of poor women around him in Bangladesh, saw an opportunity to improve their lives by giving them access to less expensive loans, unleashing a generation of microentrepreneurs. But that up-close intuition needs to be cross-checked with some statistical rigour. The microcredit schemes that Yunus did so much to popularise have now been examined more thoroughly, using randomised trials in which a group of otherwise similar people applying for small loans are either approved, or rejected, at random. (This is like a clinical trial in which some patients get a new drug while others get a placebo.) These experiments tend to find that the benefits of receiving a small loan are quite modest, and temporary. Apply the same rigorous test to other approaches – for example, giving microentrepreneurs small cash payments along with advice from a mentor – and you find that the cash-and-mentor scheme is more likely to boost the income from these tiny businesses than providing loans.[14]

---

* Admirers of Daniel Kahneman and his book *Thinking, Fast and Slow* may recognise what he calls 'system 1' and 'system 2' here.

Statistical evidence can feel dry and thin. It doesn't touch us in the same memorable and instinctive way as our personal experience. Yet our personal experience is limited. My trip to China took in tourist spots, airports and high-speed rail links. It would be a serious mistake to believe I saw everything that mattered.

There is no easy answer to the balance between the bird's eye view and the worm's eye view, between the broad and rigorous but dry insight we get from the numbers and the rich but parochial lessons we learn from experience. We must simply keep reminding ourselves what we're learning and what we might be missing. In statistics, as elsewhere, hard logic and personal impressions work best when they reinforce and correct each other. Ideally we'll find a way to combine the best of both.

One effort to do that has been developed by Anna Rosling Rönnlund of Gapminder, a Swedish foundation that fights misconceptions about global development. She aims to close the gap between fast and slow statistics – between the worm's eye view and the bird's eye view – using an ingenious website, 'Dollar Street'.

On Dollar Street you can compare the life of the Butoyi family in Makamba, Burundi, with the Bi family from Yunnan, China. Imelda Butoyi is a farmer. She and her four children get by on $27 a month. Bi Hua and Yue Hen are both entrepreneurs. Their family enjoys an income of $10,000 a month. It's no surprise that life on $27 a month is very different from life on $10,000 a month. But the numbers alone don't convey the difference in a way that we can intuitively feel, or compare to our own lives.

Dollar Street attempts to fix that, as far as is possible through the medium of a computer screen, by presenting short films and thousands of photographs of different rooms and everyday objects – a cooking stove; a source of light; a toy; somewhere to store salt; a phone; a bed. In each home about 150 photographs are taken of these everyday places and things – if they exist – and they're portrayed in the same way as far as is possible. The images speak with great clarity.

The photographs of Imelda Butoyi's home give a much more vivid impression than the precise-yet-thin statistic that she makes $27 a month. The house has mud walls, and a roof made of straw and mud. Light comes from an open fire. The toilet is a plank over a hole in the ground outside. The floor is packed earth. The children's toys? There are just a couple of picture books.

The Bi family home, in contrast, boasts a modern shower, a flush lavatory, a fancy hi-fi and a flat-screen TV. Their car is out front. The photographs show everything clearly, including the fact that the kitchen is surprisingly cramped, with just a couple of electric hobs for cooking.

'We can use photos as data,' says Rosling Rönnlund.[15] What makes them useful data rather than random and potentially misleading is that they're sortable, comparable, and connected to the numbers. The site allows you to filter so that you see only photographs of low-, middle- or high-income households. Or only photographs from a particular country. Or only photographs of a particular item – such as toothpaste or toys.

It's easy, for example, to look at all the images of cooking from very poor households and see that the standard method around the world is an iron pot hanging over an open fire. Wealthier households all use push-button appliances delivering controllable electricity or gas. Regardless of where you live, if you're poor you're likely to sleep on the floor in the same room as other family members. If you're rich you'll have privacy and a comfortable bed. Much of what we think of as cultural differences turn out to be differences in income.

'Numbers will never tell the full story of what life on Earth is all about,' wrote Hans Rosling, despite being the world's most famous statistical guru. (Hans was Anna Rosling Rönnlund's father-in-law.) Hans was right, of course. Numbers will never tell the full story – which is why, as a doctor and academic, he travelled so widely, and why he so expertly wove stories to go alongside his statistical evidence. But the stories the numbers do tell matter.

What I love about Dollar Street is that it successfully combines statistics, fast and slow – the worm's eye view and the bird's eye view. It shows us everyday images that we instinctively understand and remember. We empathise with people all round the world. But we do so in a clear statistical context – one that can show us life at $27 a month, or $500 a month, or $10,000 a month, and can make it clear how many people live in each situation.

If we don't understand the statistics, we're likely to be badly mistaken about the way the world is. It is all too easy to convince ourselves that whatever we've seen with our own eyes is the whole truth; it isn't. Understanding causation is tough even with good statistics, but hopeless without them.

And yet, if we understand only the statistics, we understand little. We need to be curious about the world that we see, hear, touch and smell as well as the world we can examine through a spreadsheet.

My second piece of advice, then, is to try to take both perspectives – the worm's eye view as well as the bird's eye view. They will usually show you something different, and they will sometimes pose a puzzle: how could both views be true? That should be the beginning of an investigation. Sometimes the statistics will be misleading, sometimes it will be our own eyes that deceive us, and sometimes the apparent contradiction can be resolved once we get a handle on what is happening. Often that will require us to ask a few smart questions – including the question I'll introduce in the next chapter.

RULE THREE

Avoid premature enumeration

Once you know what the question actually is,
you'll know what the answer means.

—DEEP THOUGHT (a supercomputer in Douglas Adams's *Hitchhiker's Guide to the Galaxy*)

It was a vital question. Across the UK, mortality rates for newborn babies varied substantially for no obvious reason. Could doctors and nurses be doing anything different to save these children? Clinicians were despatched to hospitals with better performances, instructed to think about the lessons that could be learned and to contemplate reconfiguring their own maternity services from the ground up.

But Dr Lucy Smith of the University of Leicester had a nagging doubt.[1] So she looked in detail at the data from two hospital groups, one in the English Midlands and one in London. The hospitals served very similar communities, and yet the death rates of newborns were noticeably lower in London. Were the London hospitals really doing something different in their clinics, or labour wards, or neonatal intensive care units?

No, found Dr Smith. The explanation of the disparity in mortality rates was quite different.

When a pregnancy ends at, say, twelve or thirteen weeks, everyone would call that a miscarriage. When a baby is born prematurely at twenty-four weeks or later, UK law requires this to be recorded as a birth. But when a pregnancy ends just before this cut-off point – say, at twenty-two or twenty-three weeks – how it should be described is more ambiguous. A foetus born at this stage is tiny, about the size of an adult's hand. It is unlikely to survive. Many doctors call this heart-breaking situation a 'late miscarriage', or a 'late foetal loss', even if the tiny child briefly had a heartbeat or took a few breaths. Dr Smith tells me that parents who have been through this experience often feel strongly that the word 'miscarriage' is inadequate. Perhaps in the hope of helping these parents to process their grief, the community of neonatal doctors in the Midlands had developed the custom of describing the same tragedy in a different way: the baby was born alive, but died shortly after.

Mercifully few pregnancies end at twenty-two or twenty-three weeks. But after doing some simple arithmetic, Lucy Smith realised that the difference in how these births were treated statistically was enough to explain the overall gap in newborn mortality between the two hospital trusts. Newborns were no more likely to survive in London after all. It wasn't a difference in reality, but a difference in how that reality was being recorded.

The same difference affects comparisons between countries. The United States has a notoriously high infant mortality rate for a rich country – 6.1 deaths per thousand live births in 2010. In Finland, by comparison, it is just 2.3. But it turns out that physicians in America, like those in the UK's Midlands, seem to be far more likely to record a pregnancy that ends at twenty-two weeks as a live birth, followed by an early death, than as a late miscarriage. Perhaps this is for cultural reasons, or perhaps it reflects different legal or financial considerations. Whatever the reason, some – by no means all – of the high infant mortality rate in the United States seems to be the result of recording births before twenty-four weeks as live when in other countries they would be recorded as miscarried pregnancies. Looking only at babies born after twenty-four weeks, the US infant mortality rate falls from 6.1 to 4.2 deaths per thousand live births. The rate in Finland barely shifts, from 2.3 to 2.1.[2]

The issue also arises when comparing trends over time in the same country. When the infant mortality rate rose between 2015 and 2016 in England and Wales, against a history of steadily falling rates, the press understandably raised the alarm. 'Obesity, poverty, smoking and a shortage of midwives could all be factors, say health professionals,' said the *Guardian* newspaper.[3]

Indeed they could. But a group of doctors, writing to the *British Medical Journal*, pointed out that official statistics were also recording a dramatic rise in the number of live births at twenty-two weeks of gestation, or even earlier.[4] More and more doctors, it seems, were following the Midlands trend of changing their recording practices to record live births and early deaths, rather than late miscarriages. And this was sufficient to explain the increase in the infant mortality statistics.

There is an important lesson here. Often, looking for an explanation really means looking for someone to blame. The infant mortality rate is rising – are politicians not providing enough money for the health service, or is the problem caused by mothers smoking or getting fat? The infant mortality rate is lower in London than in the Midlands – what are hospitals in the Midlands doing wrong? In truth, there may never have been anything to blame anybody for at all.

When we are trying to understand a statistical claim – any statistical claim – we need to start by asking ourselves what the claim actually means.

Measuring infant mortality, at first glance, means doing something sad and simple: counting the babies who died. But think about it for a moment and you realise that the distinction between a baby and a foetus is anything but simple – it's a deep ethical question that underlies one of the most acrimonious divides in US politics. The statisticians have to draw the line somewhere. If we want to understand what is going on, we need to understand where they drew it.

The coronavirus pandemic has raised similar questions. As I write these words, on 9 April 2020, the media are reporting that in the last twenty-four hours, 887 people died with Covid-19 on the British mainland – but I happen to know that number is wrong. Data detective work from the Scottish statistician Sheila Bird tells me that the true figure is more likely to be about 1500.[5] Why such a huge disparity? Partly because some people died at home, and the statistics represent only those who died in a hospital. But mostly because these overstretched hospitals are reporting deaths with a delay of several days. Deaths announced today, a Thursday, probably took place on Sunday or Monday. And since the death toll has been growing exponentially, telling us about what happened three days ago understates how bad things are now.*

The whole discipline of statistics is built on measuring or counting things. Michael Blastland, co-creator of *More or*

* Then there is the question of what a Covid-19 death is: some who succumb were already terminally ill; some, indeed, died *with* the virus but not *of* it. With that in mind, perhaps 1500 deaths is an overstatement after all.

Less, imagines looking at two sheep in a field. How many sheep in the field? Two, of course. Except that one of the sheep isn't a sheep, it's a lamb. And the other sheep is heavily pregnant – in fact, she's in labour, about to give birth at any moment. How many sheep again? One? Two? Two and a half? Counting to three just got difficult. Whether we're talking about the number of nurses employed by a hospital (do two part-time nurses count as two nurses, or just one?) or the wealth of the super-rich (is that the wealth they declare to the taxman, or is there a way to estimate hidden assets too?) it is important to understand what is being measured or counted, and how.

It is surprising how rarely we do this. Over the years, as I found myself trying to lead people out of statistical mazes week after week, I came to realise that many of the problems I encountered were because people had taken a wrong turn right at the start. They had dived into the mathematics of a statistical claim – asking about sampling errors and margins of error, debating if the number is rising or falling, believing, doubting, analysing, dissecting – without taking the time to understand the first and most obvious fact: what is being measured, or counted? What definition is being used?

Yet while this pitfall is common, it doesn't seem to have acquired a name. My suggestion is 'premature enumeration'.

It's a frequent topic of conversation with my wife. The radio that sits on top of the refrigerator will carry some statistical claim into our home over breakfast – a political soundbite, or the dramatic conclusion of some research. For example, 'A new study shows that children who play violent video games are more likely to be violent in reality.' Despite having known my limitations for twenty years, my wife can't quite rid herself of the illusion that I have a huge spreadsheet in my head, full of every statistic in creation. So she will turn to me and ask, 'Is that true?' Very occasionally I happen to have recently researched the issue and know the answer, but far more often I can only reply, 'It all depends on what they mean . . .'

I'm not trying to model some radical philosophical scepticism – or annoy my wife. I'm just pointing out that I don't fully understand what the claim means, so I am hardly in a position (yet) to know whether it might be true. For example, what is meant by a 'violent video game'? Does Pac-Man count? Pac-Man commits heinous acts, notably swallowing sentient creatures alive. Or what about Space Invaders? There's nothing to do in Space Invaders but shoot and avoid being shot. But perhaps that is not quite what the researchers meant. Until I know what they *did* mean, I don't know much.

And how about 'play'; what does that mean? Perhaps the researchers had children* fill in questionnaires to identify those who play violent games for many hours in a typical week. Or perhaps they recruited some experimental subjects to play a game for twenty minutes in a laboratory, then did some kind of test to see if they'd become more 'violent in reality' – and how is that defined, anyway?

'Many studies won't measure violence,' says Rebecca Goldin, a mathematician and director of the statistical literacy project STATS.[6] 'They'll measure something else such as aggressive behaviour.' And aggressive behaviour itself is not easy to measure because it is not easy to define. One influential study of video games – I promise I'm not making this up – measured aggressive behaviour by inviting people to add hot sauce to a drink that someone else would consume. This 'hot sauce paradigm' was described as a 'direct and unambiguous' assessment of aggression.[7] I am not a social psychologist,

* And by 'children', do we mean five-year-olds? Ten-year-olds? Sixteen-year-olds?

so perhaps that's reasonable. Perhaps. But clearly, like 'baby' or 'sheep' or 'nurse', apparently common-sense words such as 'violent' and 'play' can hide a lot of wiggle room.

We should apply the same scrutiny to policy proposals as we do to factual claims about the world. We all know that politicians like to be strategically vague. They will often trumpet the merits of 'fairness' or 'progress' or 'opportunity', or say, in the most infuriating tic of all, 'we're proposing this policy because we think it's the right thing to do'. But even specific-sounding policies can end up meaning very little if we don't understand the claim. You'd like to increase funding for schools? Great! Is that a funding increase per pupil, after inflation – or not?

For example, a policy paper published in the UK in 2017 by the Brexit lobby group Leave Means Leave called for a 'five-year freeze on unskilled immigration'.[8] Is that a good idea? Hard to say until we know what the idea really is: by now, we should know to ask, 'What do you mean by "unskilled"?' The answer, on closer inspection, is that you're unskilled if you don't have a job offer on a salary of at least £35,000 – a level that would rule out the majority of nurses, primary school teachers, technicians, paralegals and chemists. Now that might be a good policy or it might be a bad policy, but most people would be surprised to hear that this freeze on 'unskilled immigration' is a policy that proposes excluding people coming to work as teachers and intensive care nurses.[9] This wasn't just a policy paper, either: in February 2020, the UK government announced new immigration restrictions using a lower cut-off (a salary of £25,600) but similar language about 'skilled' and 'unskilled'.[10]

Premature enumeration is an equal-opportunity blunder: the most numerate among us may be just as much at risk as those who find their heads spinning at the first mention of a fraction. Indeed, if you're confident with numbers you may be more prone than most to slicing and dicing, correlating and regressing, normalising and rebasing, effortlessly manipulating the numbers on the spreadsheet or in the statistical package – without ever realising that you don't fully understand what these abstract quantities refer to. Arguably this temptation lay at the root of the last financial crisis: the sophistication of mathematical risk models obscured the question of how, exactly, risks were being measured, and whether those measurements were something you'd really want to bet your global banking system on.

Working on *More or Less*, I found the problem everywhere. After working with a particular definition for years, the experts we talked to could easily forget that the ordinary listener might have something very different in mind when they heard the term. What the psychologist Steven Pinker calls the 'curse of knowledge' is a constant obstacle to clear communication: once you know a subject fairly well, it is enormously difficult to put yourself in the position of someone who doesn't know it. My colleagues and I weren't immune. When we started researching some statistical confusion, we'd habitually start by pinning down the definitions – but as we quickly took them for granted, we always had to remind ourselves to explain them to our listeners, too.

Darrell Huff would be quick to point to the fact that an easy way to 'lie with statistics' is to use a misleading definition. But we can often mislead ourselves.

Consider the number 39,773. That was the number of gun deaths in the United States in 2017 (this number is from the National Safety Council and is the most recent available from that source). This number, or something very like it, is repeated every time a mass shooting makes the headlines, even though the vast majority of these deaths are nothing to do with these grim spectacles.* (Not every mass shooting is headline news, of course. Using the common definition of four people killed or injured in a single incident, there is a mass shooting almost every day in the United States, and many of them would be well down the news editor's order of priorities.)

'Gun death' doesn't sound like a complicated concept: a gun is a gun and dead is dead. Then again, nor does 'sheep', so we should pause to check our intuitions. Even the year of death, 2017, isn't as straightforward as you might think. For example, in the UK in 2016, the homicide rate rose sharply. This was because an official inquest finally ruled that ninety-six people who died in a crush at the Hillsborough football stadium in 1989 had been unlawfully killed. Initially seen as accidental, those deaths officially became homicides in 2016. This is an extreme example, but there are often delays between when somebody died and when the cause of death was officially registered.

But the big question here is about the connotations of 'death'. True, it's not an ambiguous concept. But we hear the number '39,773' at the very moment we are watching news footage showing lines of ambulances and police cars at the sight of some vivid and horrifying slaughter. So we naturally associate it with murder, or even mass murder. In fact, about 60 per cent of gun deaths in the United States are suicides, not homicides or rare accidents. Nobody set out to mislead us

* Even the definition of 'mass shooting' is slippery. The FBI keeps a record of incidents of mass murder, but their definition only includes attacks in a public place, which leaves out many drug-related incidents, as well as domestic homicides. An alternative count, maintained by the Gun Violence Archive, includes such incidents. That makes a big difference to the total count – but either way, the number of people killed in mass shooting incidents is a small fraction of the total number of gun deaths.

into thinking gun-related homicides are two and a half times more common than they actually are. It's just an assumption we understandably make from the context in which we are usually presented with the number.

Having noticed our error, what conclusions we should draw from it is another question. It's possible to spin it to support various political outlooks. Gun rights advocates will claim that it shows the fear of mass shootings is overblown. Gun control advocates will counter-claim that it weakens a common argument of the gun rights lobby – that people should be able to arm themselves to defend against an armed attacker, which is no help if the bigger risk is that people will turn their guns against themselves.

As thoughtful readers of statistics, we don't need to rush to judgement either way. Clarity should come first; advocacy can come once we understand the facts.

We should also remember that behind every one of those 39,773 gun deaths is a tragic human story. There's little evidence that Stalin ever said 'The death of one man is a tragedy, the death of millions is just a statistic', but the aphorism has echoed down the years in part because it speaks to our profound lack of curiosity at the human stories behind the numbers. Premature enumeration is not just an intellectual failure. Not asking what a statistic actually means is a failure of empathy, too.

Staying with the grim subject of suicide, this time in the UK: 'A Fifth of 17- to 19-year-old Girls Self-harm or Attempt Suicide' blares a headline in the *Guardian*. The article goes on to speculate that this may be because of social media, pressure to look good, sexual violence, pressure to do well in exams, difficulty finding work, moving to a new area, cuts in central government services, or iPads.[11] But while the piece is long on scapegoats, it's short on detail about what is meant by self-harm.

So let's turn to the original study, funded by the UK government and conducted by some respected research organisations.[12] It doesn't take long to realise that an error has slipped into the headline, as errors often do. It's not true that a fifth of seventeen- to nineteen-year-old girls self-harm or attempt suicide. What is true is that a fifth of them say that they have done so at some stage – not necessarily recently. But . . . 'done so'. What exactly have they done? The study itself is no more illuminating than the *Guardian* report on it.

The National Health Service website lists a variety of self-harming behaviours, including cutting or burning your skin, punching or slapping yourself, eating or drinking poisons, taking drugs, misusing alcohol, eating disorders such as anorexia and bulimia, pulling out your hair, or even excessive exercise.[13] Is this what these young women were thinking of when they answered 'yes' to the question? We don't know. I asked the researchers what their question meant; they told me they wanted 'to capture the entire spectrum of self harm' and so did not provide a definition of self-harm to the young women they interviewed. Self-harm means whatever the interviewees thought it means.[14]

That's fine; there is nothing necessarily wrong with aiming to capture the broadest possible range of behaviour. It might be useful to know that a fifth of seventeen- to nineteen-year-old girls have at some point behaved in a way they subjectively consider to be self-harm. But those of us interpreting the statistic will want to bear in mind that nobody else can know precisely what they meant. All forms of self-harm are disturbing, but you may find some of them a great deal more disturbing than others. Binge drinking seems very different from anorexia.

Bearing this in mind, the headline lumping together self-harm and suicide, which at first glance seemed natural, starts to look irresponsible. There is an enormous gulf between excessive exercise and killing yourself. And while this survey suggests that self-harm is worryingly common among young women, suicide is thankfully quite rare. Out of every 100,000 girls in the UK aged between fifteen and nineteen, 3.5 kill themselves each year; that's about seventy across the entire country.[15]

(By now I hope you are wondering what exactly the authorities mean by 'suicide'. It is not always clear whether someone intended to kill themselves; sometimes people intended only to hurt themselves but died by accident. In the UK, the Office for National Statistics draws a clear line: if the child is fifteen or over, the death is assumed to be deliberate; under the age of fifteen, it is assumed to be an accident. Evidently, those assumptions will not always reflect the truth, which is sometimes impossible to know.)

Lumping together self-harm and suicide is all the more irresponsible because the headline singles out girls. The study did indeed find that seventeen- to nineteen-year-old girls are much more likely than seventeen- to nineteen-year-old boys to say they had harmed themselves, yet it is the boys who are the bigger suicide risk. Boys of this age are twice as likely as girls to kill themselves.

Awful tragedies lie behind each of these numbers. Pinning down the definitions is vital if we want to understand what is happening and, perhaps, how we might make life better. That is, after all, why we're collecting the numbers.

I'd like to devote the rest of the chapter to a more detailed example, which I hope will illustrate how we might try to think through a complex problem – first by clarifying what's being measured, and only then by breaking out the mathematics. It's an important issue, but also an issue about which many people have very strong beliefs, yet a weak grasp of the definitions involved. That issue is inequality. Let's start with perhaps the most famous soundbite on the topic.

'Oxfam: 85 Richest People as Wealthy as Poorest Half of the World'. That was a *Guardian* headline in January 2014.[16] The *Independent* picked up on the same research published by the development charity Oxfam, as did many other media outlets. It's an astonishing claim. But what does it tell us?

Oxfam's aim was publicity. They wanted to generate heat; if they shed any light on the subject, that was a secondary consideration. This isn't just my opinion: the report's lead author, Ricardo Fuentes, said as much when interviewed for an Oxfam blog post titled 'Anatomy of a Killer Fact', which celebrated the 'biggest-ever traffic day on the Oxfam International website'.[17] The blog post focuses on all the attention the claim received. But was the 'Killer Fact' informative, or even true? Mr Fuentes later told the BBC that his research 'has shortcomings but it was as good as it gets'.

I'm not so sure about that. Three years later, Oxfam had revised its analysis so comprehensively that the headline number had changed from eighty-five billionaires to eight billionaires. Had the inequality really become ten times worse, the billionaires ten times richer – or perhaps the poor of the world had lost nine tenths of their wealth somehow? No, there was no such economic cataclysm. Oxfam's measure was just a very noisy and uninformative way to think about inequality in the first place.

The dramatic change in the headline claim is one indication that this may not be a terribly educational way to think about inequality. The excited bewilderment of some of the media reporting is another sign of just how baffling the number really was. While the *Guardian* accurately repeated Oxfam's headline – eighty-five people among them have as much wealth as the poorest half of the world – the *Independent* published an infographic declaring that the eighty-five richest people had as much wealth as the rest of the world put together. (A trailer for a BBC documentary about the super-rich repeated the error.) That's not even *close* to being the same claim, although you might have to think twice about why.

If thinking twice didn't help: almost all global wealth is held neither by the poorest half of the world, who have little or nothing, nor by the richest eighty-five (or eight?) ultra-billionaires. It lies with a few hundred million prosperous people in the middle. You may very well be one of them. The *Independent* and the BBC had mixed up 'the wealth of the poorest half' and 'the wealth of everyone who isn't a zillionaire'. This apparently minor confusion turns out to be between a sum of less than $2 trillion and a sum of more than $200 trillion. Not thinking hard enough about the exact claim being made introduced a hundred-fold error.

In a magnificent display of statistical befuddlement, the *Independent* also declared 'The 85 richest people – 1%' to have the same wealth as 'Rest of the world – 99%'. This implies that the population of the world is 8500. If the previous claim was a hundred-fold error, this one is nearly a million times too small.

The hopeless confusions of the *Independent* are worth dwelling on for a moment. They remind us how easy it is for our emotions to run away with us. There are some people out there with extraordinary, imagination-boggling fortunes. There are other people out there with nothing. It's not fair. And as we start to seethe at the unfairness, the risk is that we stop thinking. The *Independent* confused nearly 8 billion people with 8500 people. It confused the wealth of the poorest half of the world with the wealth of *everyone* except the richest eighty-five people. These are ludicrous errors – but as Abraham Bredius showed us, when we stop thinking and start feeling, ludicrous errors show up very promptly.

It's a nice little reminder to all of us to stop and think for a moment. It should not be too complicated a calculation to realise that whoever 'the 1%' might be, there are more than eighty-five of them.

I can't blame Oxfam, an organisation devoted to campaigning and fundraising, for seeking the most sensational headlines possible. Nor do I hold them responsible for the fact that the claim prompted all kinds of screw-ups from the media.

The rest of us, however, might prefer some clarity. So – back to the drawing board, and that starts with being clear about what's being measured, and how.

What's being measured is net wealth – that is, assets such as houses, shares and cash in the bank, less any debts. If you have a house worth $250,000 with a $100,000 mortgage on it, that's $150,000 of net wealth.

The Oxfam calculations on which the headline was based took the best available estimate of the total net wealth of the poorest half of the world (accumulated by researchers paid by a bank, Credit Suisse)[18] and compared it to the best available estimate of the total wealth of the top multi-billionaires (as reported by newspaper rich lists). They found that you only had to total up the wealth of the eighty-five richest billionaires before you exceeded the total wealth of the poorest half of the world, about 2.4 billion adults (Credit Suisse's researchers ignored children).

But does net wealth really tell us much? Let's say you buy

a nice $50,000 sports car with a $50,000 loan. The moment you drive it off the lot, the sports car has lost a few thousand dollars in value, and your net wealth has just fallen. If you've just finished an MBA, or law school, or medical school, and you've picked up a few hundred thousand dollars of debt, your net wealth is *way* below zero. But financially, a young doctor is likely to feel much more comfortable than a young subsistence farmer, even if the doctor is up to her chin in debt and the farmer owns a scrawny cow and a rusty bike for a net worth of $100.*

Net wealth is a great way to measure riches, but not such a good way to measure poverty. Lots of people have zero, or less than zero. Some of them are destitute; others, like the junior doctor, are going to be fine.

A further problem is that when you add up all those zeros and negative numbers, you're never going to get a positive number. As a result, my young son's piggy bank is worth more than the assets of the poorest billion people in the world put together, because a billion zeros and negative numbers never get you above the £12.73 he had in there when we last counted it all up. Does that suggest that my son is rich? No. Does it demonstrate that grinding poverty is endemic? Well, no, not directly. The fact that more than a billion people have no wealth is striking, but it's not clear that trying to add up all those zeros tells us much more. I'm not sure that it tells us anything, except that a billion times zero is zero.

Now that we've avoided premature enumeration – rushing to work with the numbers before we really understand

---

* And there's an oft-repeated anecdote about Donald Trump, years before he became President and heavily indebted after some failing real-estate deals, pointing to a homeless person and telling his young daughter, 'See that bum? He has a billion dollars more than me.' I've no idea if the story is true, but the financial logic is sound.

what those numbers are supposed to mean – it's the perfect time for a little light mathematics, which can be wonderfully clarifying.

Looking at the *Global Wealth Report* from Credit Suisse, the source of Oxfam's claims, we can play with some of those numbers to shed more light on the topic.*

- 42 million people have more than a million dollars each, collectively owning about $142 trillion. A few of them are billionaires, but most are not. If you have a nice house with no mortgage, in a place such as London, New York or Tokyo, that might easily be enough to put you in this group. So would the right to a good private pension.†[19] Nearly 1 per cent of the world's adult population are in this group.
- 436 million people, with more than $100,000 but less than a million, collectively own another $125 trillion. Nearly 10 per cent of the world's adult population are in this second group.
- Those two groups, collectively, have most of the cash.
- Another billion people have more than $10,000 but fewer than $100,000; they own about $4 trillion among them.
- The remaining 3.2 billion adults have only $6.2 trillion,

* I've used the 2018 *Global Wealth Report*. The 2013 version – the foundation of the original '85 Richest People' headlines – offers slightly different numbers but the big picture has changed only slowly.

* Credit Suisse did not include the entitlement to a state pension in its calculations. That matters, because state pensions are very valuable to those who have them. It's unclear whether counting state pensions as assets would increase measured inequality (since many of the poorest people lack them) or reduce measured inequality (since a state pension represents a substantial asset for the poorer people in richer countries). I'm guessing that things would look less unequal if state pensions were included, but it is just a guess. I might be quite wrong. Around the world, a third of older people have no pension of any kind.

less than $2000 each on average. Many of them have much less than that average.

Very roughly speaking, the richest half a billion people have most of the money in the world, and the next billion have the rest. The handful of eighty-five staggeringly wealthy super-billionaires are still just a handful, so they own less than 1 per cent of this total. All this, it seems to me, tells us a great deal more about the distribution of assets than a widely repeated 'killer fact' that talks about wealth inequality while ignoring almost all the wealth in the world. And while Oxfam's aim, understandably, is to produce such 'killer facts' to win attention and raise money, my aim is to understand our planet and our society. Those facts were easily accessible online; it was a matter of an extra click or two. To find them, all it took was a couple of minutes, and a curiosity about the world.

At least Oxfam was clear that it was talking about inequality of wealth. Often we hear someone make a vague assertion like 'inequality has risen' and we can't even guess that much: inequality of what, between whom, and measured how?

Perhaps they're talking about wealth inequality, having read Oxfam's stat updating the eighty-five billionaires to just eight. Or perhaps they mean inequality of income. If you want to understand how people live and what they are able to consume from day to day, inequality of income is a more natural thing to examine. What we eat, what we wear and how we live tends to be related not to our wealth but to regular income from a salary, a pension, payments from the state, or the profits from a small business. Very few people have enough wealth to fund their lifestyle purely out of interest payments, and so if we want to understand how inequality manifests itself in everyday life, it makes sense to look at income rather than wealth. The other advantage of looking at income is that we do not need to confront the absurdity of suggesting that an ordinary schoolboy and his piggy bank are richer than a billion people put together.

If we look at inequality of income, inequality between whom? The obvious answer: between the rich and the poor. But there are other possibilities: one could look at inequality between countries, or between ethnic groups, or between men and women, or between the old and the young, or between different regions within a country.

But even once we've settled on looking at inequality of income, and between high earners and low earners, the question remains: measured *how*?

Here are a couple of possibilities. You could compare the median income (the income of a person right in the middle of the distribution) to the tenth percentile income (the income of someone near the bottom of the income distribution). This is called the 50/10 ratio, and it's an indication of how the poor are doing relative to the middle class.

Alternatively, you could look at the income share of the highest-earning 1 per cent – a decent indicator not just of how the billionaires are faring, but the millionaires too. You don't need to do this yourself: think tanks and academics have done these calculations and they are usually easy to find online.[20]

Both of these measures seem to tell us something important. But what if they conflict? Imagine a country where the income of the highest-earning 1 per cent surged, while at the same time there was a reduction in inequality further down the income scale, as the 50/10 ratio shrank and poorer households caught up with the comfortably off. If the rich grow richer but the poor grow richer too, relative to the median, has inequality risen? Or fallen? Or a bit of both?

This might seem like a cute hypothetical question, but as it happens it describes the situation in the United Kingdom between 1990 and 2017. After taxes, the top 1 per cent saw their share of income rise, but inequality among lower-earning households fell as poorer households tended to catch up on those with middling incomes. It's an awkward story for anyone who wants an easy answer, but in a complicated world we shouldn't expect that the statistics will always come out neatly.

A few years ago I was invited to be the resident data geek on a TV debate about inequality in the UK. The show was an ambitious hour-long special in front of a studio audience during which various worthies would discuss why inequality in the UK mattered. In early discussions with the programme's production team, I pointed them towards the World Inequality Database, a resource that was originally put together by the economists Sir Tony Atkinson and Thomas Piketty. Piketty, of course, was the superstar author of *Capital in the Twenty-First Century*; Sir Tony, who died in 2017, was one of his academic mentors. The two of them favoured stiff redistributive taxes and wide-ranging government intervention in the economy. Like many economists, I'm quite wary of that sort of policy, but I recommended their database anyway. They were simply the world's leading experts.

All seemed well until, a few days before the show, I had an awkward phone call with one of the production team. I mentioned in passing that the pre-tax income share of the top 1 per cent had fallen slightly over the previous few years. As we've seen, that's by no means the only way to measure inequality, but it's a metric Piketty and Atkinson like to emphasise, and it seemed a good starting point: it was crisp, rigorous and easy to explain on TV. Alarmed, she told me that the entire programme was based on the premise that inequality had been increasing since the 2007–08 financial crisis.

Why did they think this was true? The data were clear: the top 1 per cent had 12 per cent and rising of pre-tax income in 2008, but the crisis knocked that back to 10 or 11 per cent.* This was hardly astonishing: a massive financial crisis is likely to temporarily hit the income of high-earners such as bankers, lawyers and corporate executives. And this was data, remember, gathered by two left-leaning economists who would have been first in line to decry the effects of bankers' greed or government cutbacks.

But no: the idea that inequality had risen just seemed to the TV producers like the kind of thing that *should* be true. Perhaps they looked at the data I'd recommended and found some flaw with it. Perhaps they found some different measure that they felt was better. But the strong impression from my conversation was that the production team simply hadn't looked at the data I'd recommended to them. I hope that isn't so, because it takes a special lack of curiosity to be able to produce an ambitious TV programme without taking the ninety seconds or so necessary to check whether the premise of the show is actually true.

I made my excuses and did not participate.

Statisticians are sometimes dismissed as bean-counters. The sneering term is misleading as well as unfair. Most of the concepts that matter in policy are not like beans; they are not merely difficult to count, but difficult to define. Once you're sure what you mean by 'bean', the bean-counting itself may come more easily. But if we don't understand the definition then there is little point in looking at the numbers. We have fooled ourselves before we have begun.

---

* Another popular measure of inequality – one we'll encounter in the next chapter – is the Gini coefficient. This measure was telling the same story of falling inequality in the wake of the crisis.

The solution, then: ask what is being counted, what stories lie behind the statistics. It is natural to think that the skills required to evaluate numbers are numerical – understanding how to compute a percentage, or to disentangle your millions from your billions from your zillions. It's a question of mathematics, is it not?

What I hope we've learned over the past few pages is that the truth is more subtle yet in some ways easier: our confusion often lies less in numbers than in words. Before we figure out whether nurses have had a pay rise, first find out what is meant by 'nurse'. Before lamenting the prevalence of self-harm in young people, stop to consider whether you know what 'self-harm' is supposed to mean. Before concluding that inequality has soared, ask 'Inequality of *what*?' Demanding a short, sharp answer to the question 'Has inequality risen?' is not only unfair, but strangely incurious. If we are curious, instead, and ask the right questions, deeper insight is within easy reach.

RULE FOUR

Step back and enjoy the view

> The shortest-lived creatures on the Disc were mayflies, which barely make it through twenty-four hours. Two of the oldest zigzagged aimlessly over the waters of a trout stream, discussing history with some younger members of the evening hatching.
>
> 'You don't get the kind of sun now that you used to get,' said one of them.
>
> 'You're right there. We had proper sun in the good old hours. It were all yellow. None of this red stuff.'
>
> 'It were higher, too.'
>
> 'It was. You're right.'
>
> —TERRY PRATCHETT, *Reaper Man*

The newspapers had an alarming message for Londoners in April 2018:

'London's Murder Rate Is Higher than New York's for the First Time Ever!' The headlines played into a narrative of gangs gone wild. And if we ignore for a moment that the very definition of 'murder' differs on either side of the Atlantic, this claim is also perfectly true. In February 2018, there were fourteen murders in New York City, but fifteen in London.[1]

But what should we conclude? Nothing.

We should conclude nothing because that pair of numbers alone tells us very little. If we want to understand what's happening, we need to step back and take in a broader perspective.

Here are a few facts worth knowing about murders in London and New York. London had 184 murders in 1990, while New York suffered 2262 – more than ten times as many. It's with that image in mind of New York as a murderous place that Londoners are alarmed at the idea that they might have become as rotten as the Big Apple. But London's murder rate has fallen, not risen, since 1990. In 2017, there were 130 murders in London, including ten people killed in terrorist attacks. London was safe in 1990 and it's a little bit safer today. As for New York, murders fell to 292 in 2017. That means New York is still more dangerous than London, but much, much safer than in 1990.

(We should really look at the murder rate per million people rather than the murder total, but the populations of New York City and London are similar, so let's not worry about that.)

Now that New York is vastly safer, very occasionally it has a good month and London has a bad one, and New York's monthly murder count dips below London's. The thing about numbers is that over time, they do tend to go up and down a bit.*

* In 2019, for example, London saw 149 murders – the highest number for a decade. There has been a rise since 2016. UK media tend to present this rise as apocalyptic; with context, it looks less worrying but is clearly a move in the wrong direction. A temporary blip, or a reversal of the long decline in murder rates? 'Only time will tell' is a cliché; it's true, though.

So while the newspaper headlines are narrowly correct, they point us away from the truth rather than towards it: the news is good, not bad; London has become safer, not more dangerous; and London remains safer than the fast-improving New York. We get the real story only with context.

In 1965, two Norwegian social scientists, Johan Galtung and Mari Ruge, made a fascinating observation: what counts as 'news' depends very much on the frequency with which we pay attention.[2] If media outlets know most of their audience is checking in every day, or every few hours, they will naturally tell us the most attention-grabbing event that's happened in that time.

Consider the financial news. There is a big difference between the rolling business coverage of Bloomberg TV, the daily rhythm of the newspaper the *Financial Times* (my employer), and the weekly take of *The Economist*, even if the three outlets have a similar interest in business, economics and geopolitics. Bloomberg might pick up on sharp market moves over the past hour. The same moves won't merit a mention in *The Economist*. Weekly, daily, hourly – the metronome of the news clock changes the very nature of what is news.

Now imagine a much slower rhythm of news: a twenty-five-year newspaper, say. What would the latest edition say? It would be packed with updates, some hopeful and some grim; it would describe the rise of China, the World Wide Web and smartphones, the emergence of al-Qaeda and the collapse of Lehman Brothers. There might be a small feature article on crime, noting that the murder count had fallen in London, but not nearly as much as in New York. Nobody would spare a syllable on the idea that London was experiencing a killing spree; such an observation could only make sense in a fast-twitch media outlet.

How about a fifty-year newspaper? Max Roser, a young economist who created the Our World in Data website, proposed that idea, inspired by Galtung and Ruge. Roser imagines a newspaper published in 1918, 1968 and 2018. Topics that seemed earth-shattering to the daily newspapers of the time might not be mentioned at all, while huge changes in the world would scream from the front pages.[3]

What would the front page of the fifty-year newspaper say in 2018? One possibility might be a story about something that didn't happen: 'Phew! World Avoids Nuclear Armageddon!' Readers of the 1968 newspaper would have read anxiously about how, over the previous three decades, the atomic bomb had been invented, developed, used on Japan with catastrophic effect, then superseded by vast arsenals of much more powerful hydrogen bombs, and how the superpowers had flirted with nuclear conflict repeatedly – in the Korean War, during the Cuban missile crisis, and more than once over Berlin. For a reader picking up a newspaper in 2018 for the first time since 1968, it would be big news that the Cold War had simply ended without a nuclear exchange of any kind – even if no daily newspaper would have been tempted in the meantime to run with a headline reading 'No H-bombs Dropped Today'.

Or perhaps the editors would splash with a story on climate change. Since early research on the greenhouse effect probably wouldn't have merited a mention in the 1968 edition, the 2018 newspaper would have to start with an explanation of the basic problem: burning fossil fuels such as gas, oil and coal turns out to alter the composition of the atmosphere in a way that helps it trap heat. (Headline: 'Gah! Burning Coal Turns Out to be a Terrible Idea!') That explanation would be illustrated by an alarming graph showing the increase in global temperatures.

Climate change is a difficult thing to report over a short time horizon. On an annual basis global temperatures bounce up and down; you can find almost as many years when they have fallen as when they have risen – which is raw material for the manufacturing of doubt. The fifty-year newspaper, however, conveys the grim news clearly: temperatures have risen by about 0.75°C since the 1960s, depending on exactly what temperature measure you look at and between which years.[4] Alas, from the right perspective, the trend is clearly that the planet is heating up.

How about a hundred-year newspaper? The perspective changes again. Thinking about readers who last consulted a newspaper in 1918, you might decide to offer a leading story about the miracle of safe childhood: 'Child Mortality Falls by a Factor of Eight!' Imagine a school set up to receive a hundred five-year-olds, randomly chosen at birth from around the world. In 1918, only sixty-eight children would have turned up for the first day of school; thirty-two would have died before reaching the age of five. This wasn't some temporary catastrophe because of the terrible 1914–18 war, or the global influenza outbreak of 1918: in 1900 the statistic would have been even worse. Now, ninety-six children show up safely for their first year in school; just four die before reaching school age. Remember, they're selected from all over the world, including the poorest, most isolated and most strife-torn of countries. That is astonishing progress.[5]

For a two-hundred-year newspaper, the editorial board might take yet another angle: 'Most People Aren't Poor!' There are still a lot of poor people, of course – between 600 and 700 million now live in what we call extreme poverty, according to the World Bank's definition as an income of less than around $1.90 per day. That's not far from one in ten of the world's population. But in the early nineteenth century, almost everyone – nineteen people out of twenty – lived in that state of destitution. That's wonderful progress, and it becomes apparent only if we step back and change our perspective.

So far I've talked about perspective mainly in terms of time. We can get useful context from other kinds of comparison, too.

Let's return to our case study of income inequality from the last chapter, where we learned there are many plausible ways to measure it – such as the 50/10 ratio, or the income share of the top 1 per cent. What if we could produce some sort of composite measure that summarises the whole of the income distribution? These composite measures exist, and we've already mentioned the most famous – the 'Gini coefficient', named after the early twentieth-century Italian statistician Corrado Gini.

Like any other measure of inequality, the Gini coefficient doesn't tell us everything. On a global scale, the coefficient has been falling – that is, incomes are becoming more equal. That's because lots of previously very poor people, many in China and India, have become a lot better off – and in the mathematical calculations that go into the Gini coefficient, that outweighs inequality rising in the upper half of the income scale, with the very rich leaving the moderately well-to-do in their wake.[6] No single number could communicate all that. But the Gini coefficient does elegantly reflect the experience of everyone across the income spectrum. Moving a dollar from a billionaire to a millionaire will not change the top 1 per cent share of income, since that dollar stays in the hands of someone in the top 1 per cent. But moving a dollar from a richer person to a poorer person, no matter how rich or poor either of them might be, will reduce the Gini coefficient.

One big problem with the Gini coefficient, however, is getting an intuitive feel for what it actually means. It's easy enough to picture a country with a Gini coefficient of zero – there, everyone gets exactly the same income. Likewise, we can readily imagine a country in which the Gini coefficient is 100 per cent – there, the despotic president has cornered all the income and everyone else gets precisely nothing. But what would it be like to live in a country where the Gini coefficient of income is, say, 34 per cent?

As it happens, if you live in the UK, you can answer that question.[7] But even a specialist in income distribution would probably understand a Gini coefficient of 34 per cent only in reference to the Gini coefficients of other countries. It's 50 in China, for example, 42 in the United States, 25 in Finland. Globally, including everyone who lives in the poorest sub-Saharan nations and the richest petrostates, the Gini coefficient of income is 65 per cent, higher than in any individual nation.[8]

But we can get an even better intuitive feel for what the Gini coefficient means by doing the same calculation on things other than income. Take life itself. Like income, life is unequally distributed. Some babies die almost immediately after being born; others live for a hundred years. But these extremes are relatively unusual: most people live for at least sixty years, and few live for more than ninety. So we would expect the global Gini coefficient of life expectancy to be fairly low, and it is – less than 20 per cent.

How about the height of adults? We all have an intuitive sense of how little that varies, so it can be another useful reference point. If my back-of-the-envelope calculation is correct, the Gini coefficient is less than 5 per cent.

For a newspaper column, I once calculated the Gini coefficient of recent sexual activity in the UK among thirty-five- to forty-four-year-olds. I know you're curious: it's 58 per cent, much higher than the UK income Gini of 34 per cent.[9] Should we be surprised that it is higher than the income Gini? I'm not sure. But it is. It seems that a ten-fold gap in sexual activity – with one person having sex once a month, and another having sex ten times a month – is far more common than a ten-fold gap in incomes. A ten-fold gap in longevity – a centenarian and a child who dies at the age of ten – is thankfully rarer still. A ten-fold gap in adult heights? Unheard of, even in the record books.

Another way to step back and enjoy the view is to give yourself a sense of scale. Faced with a statistic, simply ask yourself, 'Is that a big number?' The creators of *More or Less*, Michael Blastland and Sir Andrew Dilnot, made a habit of asking this unassuming but powerful question.[10]

Take, for example, the claim that Donald Trump's border wall between the US and Mexico would cost $25 billion to build. Is that a big number? It certainly sounds biggish, but to really understand the number you need something to compare it with. For example, the US defence budget is a little under $700 billion, or $2 billion a day. The wall would fund about two weeks of US military operations. Or, alternatively, the wall would cost about $75 a person: there are about 325 million people in the US, and $25 billion divided by 325 million is about $75.* Big number? Small number? You can be the judge of that, but I'm guessing your judgement will be better informed having made these comparisons.

Andrew Elliott, an entrepreneur who likes the question so much he published a book with the title *Is That a Big Number?*,

* If Mexico paid for the wall, the cost per person would be almost $200, since the population of Mexico is smaller. If.

suggests that we should all carry a few 'landmark numbers' in our heads to allow easy comparison.[11] A few examples:

- The population of the United States is 325 million people. The population of the United Kingdom is 65 million. The population of the world is 7.5 billion.
- Name any particular age (under the age of sixty). There are about 800,000 people of that age in the UK. If a policy involves all three-year-olds, for example, there are 800,000 of them. In the US, there are about 4 million people of any particular age (under the age of sixty).
- Distance around the Earth: 40,000km, or 25,000 miles. It varies depending on whether you go around the poles or around the equator, but not much.
- The drive from Boston to Seattle: 5000km.
- Length of a bed: 2 metres (or 7 feet). As Elliott points out, this helps you visualise the size of a room: how many beds is that?
- The gross domestic product of the US – about $20 trillion (or $20,000 billion). It's a lot of walls, if that's really how you want to spend it.
- 100,000 words: the length of a medium-sized novel.
- 381 metres: the height of the Empire State Building. (It's also about a hundred storeys.)

Personally, I like to carry a few of these numbers around in my head. I'm a geek that way. And I find that the more landmarks I have, the more sense all the other landmarks make. But the truth is that we don't *have* to remember any of these numbers. We can look any of them up, from any number of reputable sources, using any reference book or internet connection – and in many cases it will be worth double-checking anyway.

Once we have some landmark numbers to hand, they're easy to use. You can compare one thing to another (this 10,000-word report seems long but an ordinary novel is ten times longer) or you can divide one thing by another (the US defence budget is over $2000 per American, per year). Memorise, or look up, some handy numbers, and then do some simple arithmetic – with a calculator if you want. It isn't hard. But it is remarkably illuminating.*

It would be nice if we didn't have to do this – if we could rely on the media outlets that present us with statistics also helpfully providing all the context and perspective we need to make sense of them. The better ones will indeed try to do this. But context and perspective are never going to be on the front page above the fold.

We've seen one reason for this: our frequency of engagement. The splash of a daily newspaper, the lead story on a TV bulletin and the top item on a website will all focus on the most dramatic, engaging and significant events since the typical news consumer will last have checked in a few hours previously. Some media critics believe there is another reason media outlets don't emphasise context and perspective: people are drawn to bad news. Hans Rosling, co-author of *Factfulness* and a wonderful campaigner for more realistic views of the

---

* Less illuminating is the habit of writing something along the lines of 'if the US national debt was a pile of dollar bills it would stretch all the way to space/to the moon/to the sun'. Some journalists seem to think this is a great way to put a big number into context. Is it? Generally I find myself stupider at reaching the end of such sentences. Do you know how many dollar bills there are in a pile a yard high? (About eight thousand. I had to look it up, of course. Anyone would.) Space is generally regarded as being 100 kilometres above us, the moon is nearly 400,000 kilometres away, and the sun 150 million kilometres away – so a pile that stretches to the sun is a lot bigger than one that stretches to space. By my calculations, the US national debt would be a pile of dollar bills reaching to the moon six times. Happy now? I find it much clearer to note that it is about $70,000 per US citizen.

world based on good data, calls this 'the negativity instinct'. And it's generally easier to make news seem bad if you omit the context.

I'm cautious about the idea that we're biased towards bad news, because in general we tend to be rather optimistic; psychologist Tali Sharot reckons that 80 per cent of us suffer from an 'optimism bias', systematically overestimating our longevity, our career prospects and our talents while being blind to the risk of illness, incompetence or divorce.[12] Daniel Kahneman, Nobel laureate and one of the fathers of behavioural economics, calls overconfidence 'the most significant of the cognitive biases'.[13] In many ways we humans are actually pretty positive creatures – perhaps a little too positive, sometimes.

A more plausible explanation is that we are drawn to surprising news, and surprising news is more often bad than good.[14] If media outlets had a bias merely to the negative, one might expect them to report regularly on, say, smoking-related deaths. Ten times as many US residents died from smoking-related diseases as from terrorism in September 2001, the month that saw the most deadly terrorist attack in the country's history.[15] Even a weekly magazine could honestly have noted at the end of that terrible week that cigarettes had killed more people than al-Qaeda. The newspapers ignored the deaths from cigarettes because they had a bias towards the shocking.

It's possible, of course, for shocking news to be positive. But the psychologist Steven Pinker has argued that good news tends to unfold slowly, while bad news is often more sudden.[16] That sounds right – it is, after all, quicker to knock something down than to build it. Following a thought experiment the great psychologist Amos Tversky once shared with a young Pinker,[17] imagine the best possible thing that could happen to you today. You could win the lottery, I suppose. (Would that really be good news?) There are certain other moments where something wonderful could happen: you could have been hoping for a baby after many months of fruitless trying, and finally the pregnancy test comes back positive; you might have applied for a promotion or a place at university, and you get it. But for most people, on most days, the possibility of some dramatic and surprising life improvement is fairly limited. Life is already good for many people; when life isn't good, it is likely to improve slowly rather than thanks to some sudden miracle.

But the possibility of some dramatic turn for the worse? That's easy to imagine. You, or a loved one, could be diagnosed with cancer, hit by a truck, or violently assaulted. Your house could be burgled, or it could burn down. You could be sacked from your job. You could be accused of a crime you didn't commit. You could discover that your partner is having an affair, or wants a divorce. I didn't have to think hard to reel those ideas off, and I'm sure you could add more without breaking sweat – or perhaps the cold sweat would break all too swiftly. The list of catastrophes could go on indefinitely.

So when media outlets want to grab our attention, they look for stories that are novel and unexpected over a short time horizon – and these stories are more likely to be bad than good.

The need to grab attention also skews the tactics of politicians, charities and other campaigners. They know that if they want to get into the headlines, they need to make surprising claims. For example, in May 2015 the British media published the alarming news that strokes were on the increase in middle-aged people; this conclusion was based on official statistics highlighted by the Stroke Association, whose chief executive commented, 'There is an alarming increase in the numbers of people having a stroke in working age.'[18] Fortunately, this is incorrect. Strokes are becoming rarer, thanks to improved diet, better treatment and public awareness campaigns; but those same public awareness campaigns encouraged people to present themselves at hospital at the first sign of a minor stroke. As a result, hospital admissions for strokes in younger people increased – or 'rocketed' as the Stroke Association put it – and the Stroke Association was on the story. The good news is that the incidence of stroke in the UK has for a long time been falling steadily and substantially across most age groups. But how could the Stroke Association be noticed with a story like that? And if they're not noticed, they can't raise money.

Or consider Oxfam's lament, late in 2016, that 'The highly successful fight against global poverty is being lost badly in one critical area – people's minds. A new global survey ... reveals that 87% of people around the world believe that global poverty has either stayed the same or gotten worse over the past 20 years, when the exact opposite is true – it has more than halved.'[19] This press release didn't win nearly as much attention as the one we discussed in the previous chapter, which said that eighty-five people (or was it eight?) owned as much wealth as half the world (or was it everyone else?). When the alarmist press releases get the headlines, no wonder people think the plight of the world is getting worse.

In the UK, people are not hugely worried about issues such as immigration, teenage pregnancy, crime and unemployment in their own areas – but they are profoundly anxious about these issues in the country as a whole. Similar results emerge if you ask people about their personal job situation versus their view of their country's economy: most people think that all is well for them personally but are worried about the society they live in.[20] Presumably this is because we personally experience our own localities, but rely on the news for information about the wider world. The 'negativity instinct' may not be a driver of negativity in news coverage, but it certainly seems to be a result.

In 1993, Martyn Lewis, then the most popular news anchor in the UK, argued that the media should spend more time covering good news stories.[21] He was sneered at by fellow journalists who caricatured his argument as simply a request for more cheery 'And finally . . .' stories of skateboarding dogs, slotted in at the end of a news programme to sprinkle a little sugar over the evening bulletin's bitter offerings. This was unfair;* Lewis explicitly called for substantive good news stories rather than the precursors of today's videos of cats surfing on Roombas.

'Good stories are there,' he wrote, 'made all the more memorable by their rarity.' Happily, this is precisely wrong. Since Lewis wrote this in 1993, 154,000 people have escaped from extreme poverty every day.[22] In 1980, the vaccines for illnesses such as measles, diphtheria and polio used to be given to about 20 per cent of one-year-olds. Eighty per cent missed out. Now at least 85 per cent of one-year-olds receive these vaccines.[23] Child mortality, as we've seen, has fallen dramatically. The good stories are everywhere. They are not made memorable by their rarity; they are made forgettable by their ubiquity. Good things happen so often that they cannot seriously be considered for inclusion in a newspaper. 'An Estimated 154,000 People Escaped from Poverty Yesterday!' True; but not news.

We don't have daily updates on how many people escape

* It was also understandable. Lewis was the author of books such as *Cats in the News* and *Dogs in the News*.

poverty, and perhaps we never will; and when I worked for the World Bank in 2004–05, we were still updating our estimates of extreme poverty only once every three years. If a newspaper decided to pick up on the story, fine, but that would be just a single story once every thousand days. No self-respecting newspaper would republish the story regularly to remind its readers, 'Not news, but still true!' So the fall in the most extreme form of poverty – and dozens of other true stories we could tell about improved literacy, democracy, votes for women, education for girls, access to clean water, immunisations, agricultural yields, infant mortality, the price of solar power, the number of deaths in plane crashes or the prevalence of hunger – goes unreported.[24]

It's not just because it's a happy story; it's because the news comes at the wrong frequency. Gloomy stories that come at the wrong frequency are often ignored too, as we've seen with smoking, the world's most persistent, and thus most boring, cause of mass fatalities. Climate change is not ignored, but it is rarely reported directly; instead, the news covers deliberate attempts to get attention for it, such as protests, summits, and the occasional scientific or government report. We see it mentioned, infuriatingly, alongside reports on the weather – but we rarely see reports on slow-moving indicators such as the world's rising temperature.

A third example is in finance. In 2004 and 2005, my *Financial Times* colleague Gillian Tett highlighted the development of huge financial markets in debt and derivatives, a kind of side-bet on the movements of interest rates, exchange rates or other financial indicators. The world financial system was like an iceberg: above the surface glistened the stock markets, easy to see and to discuss; beneath the waves lurked the debt and derivatives markets, vast and hidden. Stock markets publish numbers continually, including a daily close-of-market update for the evening news. But one of the most important measures of the size of the derivatives market is produced by the Bank for International Settlements once every three years. The pace of information didn't fit the frequency of the financial newspapers, and so it was systematically under-reported. Of course, this was bad news worth being aware of: problems in these markets were at the centre of the catastrophic financial crisis of 2007–08, and Gillian Tett was one of the few people who could honestly say she'd been paying attention beforehand.[25]

Some commentators argue that the cure for all this is simply to stop reading the newspapers. The author Rolf Dobelli – amusingly, writing in the *Guardian* newspaper – gives us ten reasons to stop reading the news.[26] Nassim Taleb, author of *The Black Swan*, puts it succinctly: 'To be completely cured of newspapers, spend a year reading the previous week's newspapers.'[27]

As someone who works for a newspaper, you might expect me to protest. I have a lot of sympathy, though. I often find that my Saturday *Financial Times* column is unmoored from the news of the week. I'm just not very interested in producing a hot take on recent news; I find my interest far more engaged by topics that have occurred to me after reading books or academic papers – or just musing about life. And while I enjoy the way that fans of *More or Less* often compare it favourably to rolling radio and TV news, I sometimes feel that we're getting credit for something that comes naturally: we operate at a different rhythm than the rolling news. As a weekly programme we usually have a couple of days to chew over something that has been said – or missed – in the blur of a live interview. Often, we find ourselves pondering a topic for weeks or months. Why cover a story quickly when you can explore it properly? And we don't usually have to worry about being scooped because we're far too nerdy for anyone else to care about our stories.

Professionally, I can't ignore the news, but I pay less attention to it than many of my colleagues – occasionally to their frustration. Daily news always seems more informative than rolling news; weekly news is typically more informative than daily news. A book is often better still. Even within a daily or a weekly newspaper, I find myself preferring the slower-paced explanation and analysis rather than the breaking news.

If you're a news junkie I suggest that you go deeper and broader, rather than faster and faster. It is harder to do this when the news itself seems to be alarming, but it's a good habit. Very little news requires the immediate attention that you might devote to a traffic update or a severe weather warning. If you come back in an hour – or a week – you will learn just as much. Indeed, you'll probably learn more. You might even ask yourself: what would a weekly magazine or a weekly podcast be paying attention to that might otherwise be drowned out in the noise of rolling news?

In the crazy early days when the Covid-19 coronavirus went global, *Scientific American* admonished journalists, 'facts about this epidemic that have lasted a few days are far more reliable than the latest "facts" that have just come out, which may be erroneous or unrepresentative and thus misleading . . . a question that today can be answered only [by] informed belief may perhaps be answered with a fact tomorrow.'[28] Sound advice, and not just for journalists but for citizens too. So however much news you choose to read, make sure you spend time looking for longer-term, slower-paced information. You will notice things – good and bad – that others ignore.

What have we learned so far about how to evaluate a statistical claim? In the first chapter, I advised trying to notice your feelings about the claim; in the second chapter, constructively sense-checking the claim against your personal experience; in the third chapter, asking yourself if you really understand what the claim means. These are all simple, common-sense suggestions, and in this chapter I've added a fourth: step back and look for information that can put the claim into context. Try to get a sense of the trend. 'Another terrible crime has occurred!' is perfectly consistent with 'Overall, crime is way down'. Look for something that will give you a sense of scale, such as comparing the situation in one country to the situation in other countries, or figuring out the cost per person of some proposed government expenditure.

None of these methods is technical; anyone can use them. Together they can go a long way towards providing statistical illumination. But sometimes we need to dig a little deeper into how a statistic was produced. Let's do that now.

RULE FIVE

Get the back story

'In each human coupling, a thousand million sperm vie for a single egg. Multiply those odds by countless generations . . . it was you, only you, that emerged. To distil so specific a form from that chaos of improbability, like turning air to gold . . . that is the crowning unlikelihood . . .'

'You could say that about anybody in the world!'

'Yes. Anybody in the world . . . But the world is so full of people, so crowded with these miracles, that they become commonplace and we forget . . .'

—ALAN MOORE, *Watchmen*

A couple of decades ago, two respected psychologists, Sheena Iyengar and Mark Lepper, set up a jam-tasting stall in an upmarket store in California. Sometimes they offered six varieties of jam, at other times twenty-four; customers who tasted the jam were then offered a voucher to buy it at a discount. The bigger display with a wider array of jams attracted more customers but very few of them actually bought jam. The display that offered fewer choices inspired more sales.[1]

The counterintuitive result went viral – it hit a sweet spot. People respond better to fewer choices! It became the stuff of pop-psychology articles, books and TED talks. It was unexpected yet seemed plausible. Few people would have predicted it, and yet somehow those who heard about it felt they'd known it all along.

As an economist, this always struck me as a little strange. Economic theory predicts that people should often value extra choices, and will never be discouraged by them – but economic theory can be wrong, so that's not what was curious about the jam study.

One puzzle was that according to the study, the measured effect of offering more choice was huge: only 3 per cent of jam tasters at the twenty-four-flavour stand used their discount voucher, versus 30 per cent at the six-flavour stand. This suggests that by trimming their range, retailers could increase their sales ten-fold. Does anybody really believe that? Draeger's, the supermarket which hosted the experiment, stocked 300 varieties of jam and 250 types of mustard. They seemed to be doing fine. Had they missed a trick? Starbucks boasts of offering literally tens of thousands of combinations of frothy drink; they seem to be doing fine, too. So I wondered just how general the finding might be. Still, it was a serious experiment conducted by serious researchers. And one should always be willing to adjust one's views to fit the evidence, right?

Then I met a researcher at a conference who told me I should get in touch with a young psychologist called Benjamin Scheibehenne. I did. Scheibehenne had no reason to doubt Iyengar and Lepper's discovery that people might be demotivated when faced with lots of options. But he had observed the same facts about the world that I had – that so many successful businesses offer a cornucopia of choice. How were those facts compatible with the experiment? Scheibehenne had a theory, which was that companies were finding ways to help people navigate complex choices. That seems plausible. Perhaps it was something to do with familiarity: people often go to the supermarket planning to buy whatever they bought last time, rather than some fancy new jam. Perhaps it was the way the aisles were signposted, or choices organised to make them less bewildering. These all seem sensible things to investigate, so Scheibehenne planned to investigate them.[2]

He began by re-running the jam experiment to get a baseline from which he could start tweaking and exploring different possibilities. But he didn't get the same baseline. He didn't get the same result at all. Iyengar and Lepper had discovered that choice dramatically demotivates. When Scheibehenne tried to repeat their experiment he found no such thing. Another researcher, Rainer Greifeneder, had re-run another similar study by Iyengar and Lepper that focused on choosing between luxury chocolates, and like Scheibehenne had failed to reproduce the original 'choice is bad' result. The pair teamed up to pull together every study of the 'choice is bad' effect they could find. There were plenty, but many of them had failed to find a journal that would publish them.

When all the studies, published and unpublished, were assembled, the overall result was mixed. Offering more choices sometimes motivates and sometimes demotivates people. Published research papers were more likely to find a large effect, either positive or negative. Unpublished papers were more likely not to find an effect at all. The average effect? Zero.[3]

This is unnerving. So far we've encountered misleading claims in the context of an agenda being pushed – Oxfam drumming up publicity, a media outlet chasing clicks – or a subtle detail being overlooked, like the use of different words to describe the tragic early end of a pregnancy. When it comes to academia, we might reasonably hope that the subtle details will be spotted and the only agenda being pursued is a search for knowledge. It makes sense to tread carefully with campaigning groups or clickbait headlines, but can't we assume we're on more solid ground when we pick up an academic journal? Iyengar and Lepper were, as I've said, highly respected. Is it possible that they were just flat-out wrong? If so, how? And what should we make of the next counterintuitive finding that sweeps the science pages or the airport bookshelves?

For an answer, let's take a step sideways, and ponder the internet's most famous potato salad.

Surely there is no easier way to raise some cash than through Kickstarter? The crowdfunding website enjoyed a breakthrough moment in 2012 when the 'Pebble', an early smartwatch, raised over $10 million. In 2014, a project to make a picnic cooler raised an extraordinary $13 million. Admittedly, the 'Coolest' cooler was the Swiss army knife of cool boxes. It has a built-in USB charger, cocktail blender and loudspeakers, attracting a thundering herd of backers. The Pebble smartwatch had its revenge in 2015, as a fresh campaign raised more than $20 million for a new and better watch.

In some ways, though, Zack 'Danger' Brown's Kickstarter achievement was more impressive than any of these. He turned to Kickstarter for $10 to make some potato salad – and he raised $55,492 in what must be one of history's most lucrative expressions of hipster irony.[4]

Following Zack Brown's exploits, I wondered what exciting project I might launch on Kickstarter, looking forward to settling back to count the money as it poured in.

The same thought may have occurred to David McGregor. He was bidding for £3600 to fund a trip across Scotland, photographing its glorious scenery for a glossy book – a lovely way to fund his art, and his holiday. Jonathan Reiter had bigger ambitions. His 'BizzFit' looked to raise $35,000 to create an algorithmic matching service for employers and employees. Shannon Limeburner was also business-minded, but sought a mere $1700 to make samples of a new line of swimwear she was designing. Two brothers in Syracuse, New York, even launched a Kickstarter campaign in the hope of being paid $400 to film themselves terrifying their neighbours at Halloween.

These disparate campaigns have one thing in common: they received precisely zero support. Not one of these people was able to persuade strangers, friends, or even their own families to kick in so much as a cent.

My inspiration and source for these tales of Kickstarter failure is Silvio Lorusso, an artist and designer based in Venice. Lorusso's website, Kickended.com, searched Kickstarter for all the projects that have received absolutely no funding. (There are plenty: about 10 per cent of Kickstarter projects go nowhere at all, and fewer than 40 per cent raise enough money to hit their funding targets.)

Kickended performs an important service. It reminds us that what we see around us is not representative of the world; it is biased in systematic ways. Normally, when we talk of bias we think of a conscious ideological slant. But many biases emerge from the way the world presents some stories to us while filtering out others.

I have never read a media report or blog post about the

attempts of the young and ambitious band Stereotypical Daydream to raise $8000 on Kickstarter to record an album. ('Our band has tried many different ways of saving money to record a legitimate album in a professional studio. Unfortunately, we still have not saved enough.') It probably will not surprise you to hear that the Stereotypical Daydream Kickstarter campaign brought them zero dollars closer to their goal.

On the other hand, I've heard quite a lot about the Pebble watch, the Coolest cooler and even that potato salad. If I didn't know better, I might form unrealistic expectations about what running a Kickstarter campaign might achieve.

This isn't just about Kickstarter, of course. Such bias is everywhere. Most of the books people read are bestsellers – but most books are not bestsellers, and most book projects never become books at all. There's a similar tale to tell about music, films and business ventures.

Even cases of Covid-19 are subject to selective attention: people who feel terrible go to hospital and are tested for the disease; people who feel fine stay at home. As a result, the disease looks even more dangerous than it really is. Even though statisticians understand this problem perfectly well, there's no easy way to solve it without systematic testing. And in the early stages of the epidemic, when the most difficult policy decisions were being made, systematic testing was elusive.

There's a famous story about the mathematician Abraham Wald, asked in 1943 to advise the US air force on how to reinforce their planes. The planes were returning from sorties peppered with bullet holes in the fuselage and wings; surely those spots could use some armour plating? Wald's written response was highly technical, but the key idea is this: we only observe damage in the planes that return. What about the planes that were shot down? We rarely see damage to the engine or fuel tanks in planes that survive. That might be because those areas are rarely hit – or it might be that whenever those areas are hit, the plane is doomed. If we look only at the surviving planes – falling prey to 'survivorship bias' – we'll completely misunderstand where the real vulnerabilities are.[5]

The rabbit-hole goes deeper. Even the story about survivorship bias is an example of survivorship bias; it bears little resemblance to what Abraham Wald actually did, which was to produce a research document full of complex technical analysis. That is largely forgotten. What survives is the tale about a mathematician's flash of insight, with some vivid details added. What originally existed and what survives will rarely be the same thing.[6]

Kickended, then, provides an essential counterpoint to the breathless accounts of smash hits on Kickstarter. If successes are celebrated while failures languish out of sight (which is often the situation) then we see a very strange slice of the whole picture.

This starts to give us a clue as to what might have happened with the jam experiment. Like the Coolest cooler, it was a smash hit – but not the full story. Benjamin Scheibehenne's role was a bit like Silvio Lorusso's at Kickended: he had gone looking not just for the choice experiment that had gone viral, but for all the other experiments that had produced different results and had vanished into obscurity. When he did, he was able to reach a very different conclusion.

Bear Kickended in mind as you ponder the following story. In May 2010, a surprising paper was submitted to the *Journal of Personality and Social Psychology*. The author was Daryl Bem, a respected old hand in the field of academic psychology. What made the research paper astonishing was that it provided apparently credible statistical evidence of an utterly incredible proposition: that people could see into the future. There were nine experiments in total. In one, participants would look at a computer screen at an image of two curtains. Behind one curtain was an erotic photograph, they were told. They simply had to intuit which one. The participant would make a choice, and then – after the choice had already been made – the computer would randomly assign the photograph. If the participants' guesses were appreciably better than chance, then that was evidence of precognition. They were.[7]

In another of the experiments that Bem's research paper described, subjects were shown a list of forty-eight words and tested to see how many of the words they would remember. Then some subjects were asked to practise by re-typing all the words. Normally it would be no surprise that practice helps you remember, but in this case Bem found that the practice worked even though the memory test came first, and the practice came after.

How seriously should we take these results? Bear in mind that the research paper, 'Feeling the Future', was published in a respected academic journal after a process of peer review. The experiments it reported passed the standard statistical tests, which are designed to screen out fluke results. All this gives us some reason to believe that Bem found precognition.

There is a much better reason to believe that he did not, of course, which is that precognition would violate well-established laws of physics. Vigorous scepticism is justified. As the saying goes, extraordinary claims require extraordinary evidence.

Still, how did Bem accumulate all this publishable evidence for precognition? It's puzzling. Perhaps it's less puzzling after you connect it to the story of Kickended.

After Bem's evidence of precognition had been published

in the *Journal of Personality and Social Psychology*, several other studies were produced which followed Bem's methods. None of them found any evidence for precognition, but the journal refused to publish any of them. (It did publish a critical commentary, but that's not the same thing as publishing an experiment.) The journal's grounds for refusal were that it 'did not publish replications' – that is, once an experiment had demonstrated an effect there was no space to publish attempts to check on that effect. In theory, that might sound reasonable: who wants to read papers confirming things they already knew? In practice, it has the absurd effect of ensuring that when something you thought you knew turns out to be wrong, you won't hear about it. Bem's striking finding became the last word.[8]

But it was also the first word. Before Bem came along, I strongly doubt that any serious journal would have published research, no matter how rigorous, whose abstract read: 'We tested several hundred undergraduates to see if they could see into the future. They couldn't.'

This, then, is a survivorship bias as strong as press coverage of Kickstarter projects or trying to deduce the vulnerabilities of planes by examining only the ones whose vulnerabilities weren't fatal. Out of all the possible studies that could have been conducted, it's reasonable to guess that the journal was interested only in the ones that demonstrated precognition. This wasn't because of a bias in favour of precognition. It was because of a bias in favour of novel and surprising discoveries. Before Bem, the fact that students didn't seem to be able to see into the future was trivial and uninteresting. After Bem, the fact that students didn't seem to be able to see into the future was a not-welcome-in-this-journal replication attempt. In other words, only evidence of precognition was publishable because only evidence of precognition was surprising. Studies showing no evidence of precognition are like bombers that have been shot in the engine: no matter how often such things happen, they're not going to make it to where we can see them.

The 'choice demotivates' finding is far more credible than the 'students can see into the future' finding – but still, the jam experiment may have been subject to a similar dynamic. Imagine approaching a psychology journal before Iyengar and Lepper's breakthrough result with the following study: 'We set up stalls offering people different kinds of cheese. Sometimes the stalls had twenty-four types of cheese and sometimes just six. On the days when people were offered more types of cheese, they were a bit more likely to buy cheese.' Yawn! That's not surprising at all. Who wants to publish that? It was only when Iyengar and Lepper ran an experiment showing the opposite result that the whole thing became not only publishable, but a Coolest-cooler smash hit.

If you read only the experiments published in the *Journal of Personality and Social Psychology*, you might well conclude that people can indeed see into the future. For obvious reasons, this particular flavour of survivorship bias is called 'publication bias'. Interesting findings are published; non-findings, or failures to replicate previous findings, face a higher publication hurdle.

Bem's finding was the $55,000 potato salad – wildly atypical, and widely reported as a result. The unpublished replications would typically have been like Stereotypical Daydream's attempts to fund their album: nothing happened and nobody cared.

Except this time, somebody did care.

'The paper is beautiful,' says Brian Nosek of Daryl Bem's study. 'It follows all the rules of what one does, does it in a really beautiful way.'[9]

But as Nosek, a psychologist at the University of Virginia, understood perfectly well, if Bem followed all the rules of academic psychology and ended up seeming to demonstrate that people can see into the future, something is wrong with the rules of academic psychology.[10]

Nosek wondered what would happen if you systematically re-ran some more respected and credible psychological experiments. How many results would come out the same? He sent round an email to like-minded researchers, and with impressive speed managed to get a global network of nearly three hundred psychologists collaborating to check studies that had recently been published in one of three prestigious academic journals. While Benjamin Scheibehenne had been digging into one particular field – the link between motivation and choice – Nosek's network wanted to cast their net widely. They chose a hundred studies. How many did their replication attempts back up? Shockingly few: only thirty-nine.[11] That left Nosek and the rest of academic psychology with one big question on their hands: how on *earth* did this happen?

Part of the explanation must be publication bias. As with Daryl Bem's study, there is a systemic bias towards publishing the interesting results, and of course flukes are more likely to seem interesting than genuine discoveries.

But there's a deeper explanation. It's the reason Nosek had to reach out to so many colleagues, rather than simply get his graduate student assistants to do all the checks. Since the top journals weren't very interested in publishing replication attempts, he knew that devoting his research team full time to a replication effort might be career suicide: they simply wouldn't be able to accumulate the publications necessary to secure their future in academia. Young researchers must either 'publish or perish', because many universities and other research bodies use publication records as an objective basis for deciding who should get promotions or research grants.

This is another example of the Vietnam body count problem we met in the second chapter. Great researchers do indeed tend to publish lots of research that is widely cited by others. But once researchers are rewarded for the quantity and prominence of their research, they start looking for ways to maximise both. Perverse incentives take over. If you have a result that looks publishable but fragile, the logic of science tells you to try to disprove it. Yet the logic of academic grants and promotions tells you to publish at once, and for goodness' sake don't prod it too hard.

So not only are journals predisposed to publish surprising results, researchers facing 'publish or perish' incentives are more likely to submit surprising results that may not stand up to scrutiny.

The illusionist Derren Brown once produced undoctored film of him tossing a coin into a bowl and getting heads ten times in a row. Brown later explained the trick: the stunning sequence came only at the end of nine excruciating hours of filming, when the string of ten heads finally materialised.[12] There is a 1 in 1024 chance of getting ten heads in a row if you toss a fair coin ten times. Toss it a few thousand times and a run of ten consecutive heads is almost guaranteed. But Brown could send his stunning result off to the 'Journal of Coin Flipping', perhaps with the delicious title (suggested by the journalists Jacob Goldstein and David Kestenbaum) 'Heads Up! Coin-Flipping Bias in American Quarter Dollars Minted in 1977'.[13]

To be clear – such a research paper would be fraudulent, and nobody believes that such extreme and premeditated publication bias explains the large number of non-replicable studies that Nosek and his colleagues unveiled. But there are shades of grey.

What if 1024 researchers individually researched coin tossing, and one of them produced the stunning result of ten heads in a row? That is mathematically the same situation, but from the point of view of the astonished researcher in question, she or he would be blameless. Now it seems unlikely that so many researchers would have bothered to investigate coin-tossing – but we don't know how many people tried and failed to find precognition before Daryl Bem succeeded.

The shades of grey also apply within an individual researcher's laboratory. For example, a scientist could do a small exploratory study. If he or she found an impressive result, why not publish? But if the study fell flat, the researcher could chalk it up as a learning experience and try something else. This behaviour doesn't sound especially unreasonable to the layman, and it probably doesn't feel unreasonable to the researchers doing it – but it is publication bias nonetheless, and it means that flukes are disproportionately likely to be published.

Another possibility is that the researcher does the study, finds some promising results, but those results are not quite statistically solid enough to publish. Why not keep going, recruiting some more participants, gathering some more data, and seeing if the results firm up? Again, this doesn't seem unreasonable. What could be wrong with gathering more data? Wouldn't that just mean that the study was getting closer and closer to the truth? There's nothing wrong with doing a large study. In general, more data is better. But if data are gathered bit by bit, testing as we go, then the standard statistical tests aren't valid. Those tests assume that the data have simply been gathered, then tested – not that scientists have collected some data, tested them, and then maybe collected a bit more.

To see the problem, imagine a game of basketball is about

to be played and someone asks you a question: how convincing would a victory have to be before you feel confident saying that the winning team is better than the other team, rather than just luckier on the day? There's no right answer – after all, sometimes luck can be outrageous. But you might decide that a margin of, say, ten points at the end of the game is enough to be convincing. This is, very roughly, what the standard statistical tests do to decide whether or not an effect is deemed to be 'significant' enough to publish.

But now imagine the organiser of the basketball game stands to get a bonus if one of the teams turns out to be better – it doesn't matter which – so, without telling you, she decides that if either team is ever ahead by ten points, she'll bring the game to an early halt. And if, at the final whistle, the two teams are separated by seven, eight or nine points, she'll play overtime to see if the gap opens up to ten. After all, she's just a basket or two away from demonstrating the superiority of one of the teams!

It's obvious that would be a misuse of the test you set, but much of this kind of misuse seems to be quite common in practice.[14]

A third problem is that researchers also have choices as to how they analyse the data. Maybe the study holds up for men, but not women.* Maybe the study holds up if the researcher makes a statistical adjustment for age, or for income. Maybe there are some weird outliers and the study holds up only if they are included, or only if they are excluded.

Or maybe the scientist has a choice of different things she or he could measure. For instance, a study of how screen use

* In our basketball analogy, that's like the organiser noticing that she can get a ten-point gap if she counts the field goals and ignores the free throws. Outrageous in that setting, but in a scientific context it makes perfect sense to explore different approaches. Perfect sense – but a statistical trap if not done carefully.

affects the well-being of young people could measure both screen use and well-being in different ways. Well-being can be measured by asking people about episodes of anxiety; or it could be measured by asking people about how satisfied they are with their lives; or it could be measured by asking a young person's *parents* how they think he or she is doing. Screen time could be measured directly through a tracking app, or indirectly through a survey; or perhaps rather than 'screen time' one might want to measure 'frequency of social media use'. None of these choices is right or wrong, but – again – the standard statistical tests assume that the researcher made the choice before collecting the data, then collected data, then ran the test. If the researcher ran several tests, then made a choice, flukes are vastly more likely.

Even if the researcher ran only one test, flukes are more likely to slip through if he or she did so after gathering the data and getting a feel for how they looked. This leads to yet another kind of publication bias: if a particular way of analysing the data produces no result, and a different way produces something more intriguing, then of course the more interesting method is likely to be what is reported and then published.

Scientists sometimes call this practice 'HARKing' – HARK is an acronym for Hypothesising After Results Known. To be clear, there's nothing wrong with gathering data, poking around to find the patterns and then constructing a hypothesis. That's all part of science. But you then have to get new data to test the hypothesis. Testing a hypothesis using the numbers that helped form the hypothesis in the first place is not OK.[15]

Andrew Gelman, a statistician at Columbia University, favours the term 'the garden of forking paths', named after a short story by Jorge Luis Borges. Each decision about what data to gather and how to analyse them is akin to standing on a pathway as it forks left and right and deciding which way to go. What seems like a few simple choices can quickly multiply into a labyrinth of different possibilities. Make one combination of choices and you'll reach one conclusion; make another, equally reasonable, and you might find a very different pattern in the data.[16]

A year after Daryl Bem's result was released, three psychologists published a demonstration of just how seriously researchers could go astray using standard statistical methods combined with these apparently trivial slips and fudges.[17] The researchers, Joseph Simmons, Uri Simonsohn and Leif Nelson, 'proved' that listening to 'When I'm Sixty-Four' by the Beatles would make you nearly eighteen months younger.[18]

I know you're curious: how did they do it? The researchers collected various pieces of information from each participant, including their age, their gender, how old they felt, the age of their fathers, and the age of their mothers – along with various other almost completely irrelevant facts. They analysed every possible combination of these variables, and they also analysed the data in sets of ten participants, stopping to check for significant results each time. In the end they found that if they statistically adjusted for the fathers' ages, but not the mothers', and if they stopped after twenty participants, and if they discarded the other variables, then they could demonstrate that people who had been randomly assigned to listen to 'When I'm Sixty-Four' were substantially younger than a control group who had been randomly assigned to listen to a different song. All utter nonsense, of course – but utter nonsense that bore an eerie resemblance to research that had been published and taken seriously. Would genuine researchers ever push so far over the line from rigorous practice into rigged research? Probably not very often. But those who did would get more attention. And the majority who did not might unwittingly commit subtler versions of the same statistical sins.

The standard statistical methods are designed to exclude most chance results.[19] But a combination of publication bias and loose research practices means we can expect that mixed in with the real discoveries will be a large number of statistical accidents.

Darrell Huff's *How to Lie with Statistics* describes how publication bias can be used as a weapon by an amoral corporation more interested in money than truth. With his trademark cynicism, he mentions that a toothpaste maker can truthfully advertise that the toothpaste is wonderfully effective simply by running experiments, putting all unwelcome results 'well out of sight somewhere', and waiting until a positive result shows up.[20] That is certainly a risk – not only in advertising but also in the clinical trials that underpin potentially lucrative pharmaceutical treatments. But might accidental publication bias be an even bigger risk than weaponised publication bias?

In 2005, John Ioannidis caused a minor sensation with an article titled 'Why Most Published Research Findings Are False'. Ioannidis is a 'meta-researcher' – someone who researches the nature of research itself.* He reckoned that the cumulative effect of various apparently minor biases might mean that false results could easily outnumber the genuine ones. This was five years before the *Journal of Personality and*

---

* Ioannidis, you may recall, is also the epidemiologist who warned of a 'one-in-a-century evidence fiasco' in March 2020, as countries around the world found themselves having to respond to the coronavirus pandemic armed with very patchy data.

Social Psychology published Daryl Bem's research on precognition, which sparked Brian Nosek's replication attempt. Precognition might not exist, but Ioannidis clearly saw the crisis coming.[21]

I confess that when I first heard of Ioannidis's research, it struck me as an extraordinary piece of hyperbole. Sure, all scientific research is provisional, everyone makes mistakes, and sometimes bad papers get published – but surely it was wrong to suggest that more than half of all the empirical results out there were false? But after interviewing Scheibehenne and learning what he'd discovered about the choice literature, I started to wonder. Then, over the years, it gradually became painfully clear to me and many others who were initially sceptical that Ioannidis was on to something important.

While Bem's precognition study was understandably famous, many other surprising psychological findings had become well known to non-psychologists through books such as *Thinking, Fast and Slow* (by Nobel laureate Daniel Kahneman), *Presence* (by psychologist Amy Cuddy) and *Willpower* (by psychologist Roy Baumeister and journalist John Tierney). These findings hit the same counterintuitive sweet spot as the jam experiment: strange enough to be memorable, but plausible enough not to dismiss out of hand.

Baumeister is famous in academic psychology for studies showing that self-control seems to be a limited resource. People asked to restrain themselves by munching radishes while delicious freshly baked chocolate cookies lay within easy reach were then quicker to abandon a frustrating task later.[22] Cuddy found that asking people to adopt 'power poses' – for example, hands-on-hips like Wonder Woman – boosted their levels of testosterone and suppressed their levels of the stress hormone cortisol.[23] Kahneman described the 'priming' research of John Bargh. Young experimental subjects were asked to solve a word puzzle in which some of them were exposed to words that suggested old age, such as *bald*, *retirement*, *wrinkle*, *Florida* and *gray*. The young subjects who had not seen these particular words then set off briskly down the corridor to participate in another task; the young subjects who had, instead, been 'primed' with words suggesting old age shuffled off down the corridor at a measurably slower pace.[24]

These are extraordinary results, but as Kahneman wrote about priming research, 'Disbelief is not an option. The results are not made up, nor are they statistical flukes. You have no choice but to accept that the major conclusions of these studies are true.'

Now we realise that disbelief *is* an option. Kahneman does, too. Publication bias, and more generally the garden of forking paths, means that plenty of research that seems rigorous at first sight both to onlookers and often to the researchers themselves may instead be producing spurious conclusions. These studies – of willpower, of power-posing and of priming – have all proved difficult to replicate. In each case, the researchers have defended their original finding, but the prospect that they were all statistical accidents seems increasingly reasonable.

Daniel Kahneman himself dramatically raised the profile of the issue when he wrote an open letter to psychologists in the field warning them of a looming 'train wreck' if they could not improve the credibility of their research.[25]

The entire saga – Ioannidis's original paper, Bem's nobody-believes-this finding, the high-profile struggles to replicate Baumeister's, Cuddy's and Bargh's research, and as the coup de grâce, Nosek's discovery that (as Ioannidis had said all along) high-profile psychological studies were more likely *not* to replicate than to stand up – was sometimes described as a 'replication crisis' or a 'reproducibility crisis'.

In the light of Kickended, perhaps none of this should have been a surprise – but it is shocking nonetheless. The famous psychological results are famous not because they are the most rigorously demonstrated, but because they're *interesting.* Fluke results are far more likely to be surprising, and so far more likely to hit that Goldilocks level of counterintuitiveness (not too absurd, but not too predictable) that makes them so fascinating. The 'interestingness' filter is enormously powerful.

Little harm is done if publication bias (and survivorship bias) merely produces cute distortions in our view of the world, leading people to prepare for a job interview by finding a secluded spot to strike a Wonder Woman pose. Even if many would-be entrepreneurs are foolishly over-optimistic about their chances of raising money on Kickstarter, we all enjoy the fruits of successful new business ideas that more rational people would not have quit their jobs to pursue. And few scientists were about to embrace Daryl Bem's apparent discovery of precognition, for reasons well summarised by Ben Goldacre, an expert in evidence-based medicine: 'I wasn't very interested, for the same reasons you weren't. If humans really could see the future, we'd probably know about it already; and extraordinary claims require extraordinary evidence, rather than one-off findings.'[26]

But Ben Goldacre thinks the stakes are higher, and so do I. This bias may have serious consequences for both our money and our health.

Money first. Business writing – a field in which I confess to dabbling – is dripping with examples of survivorship bias. In my book *Adapt*, I had a little chuckle about the Tom Peters and Robert Waterman book *In Search of Excellence*, a block-busting business bestseller published in 1982, which offered management lessons gleaned from studying forty-three of the most outstanding corporations of that time. If they really were paragons of brilliant management then one might have expected their success to last. If instead they were the winners of an invisible lottery, the beneficiaries of largely random strokes of good fortune, then we would expect that the good luck would often fail to last.

Sure enough, within two years almost a third of them were in serious financial trouble. It's easy to mock Peters and Waterman – and people did – but the truth is that a healthy economy has a lot of churn in it. Corporate stars rise, and burn out. Sometimes they have lasting qualities, sometimes fleeting ones, and sometimes no qualities at all bar some luck. By all means look at the success stories and try to learn lessons, but be careful. It is easy, in Nassim Taleb's memorable phrase, to be 'fooled by randomness'.

Perhaps all such business writing is harmless: when daily data from the shop floor contradict the business-book wisdom, the shop floor will win. While the jam study became famous among the chattering classes, there is scant sign that many businesses took the 'choice is bad' finding seriously in the decisions they made about stocking their shelves. Still, one can't help suspecting that where good data are rarer, major decisions are being taken on the basis of survivor bias.

In finance, the problem may be worse. A Norwegian TV show illustrated this rather brilliantly in 2016 by organising a stock-picking competition, in which investors would buy a variety of Norwegian shares to the value of 10,000 Norwegian Krone – about $1000. The competitors were a diverse bunch: a pair of stock brokers, who confidently opined 'the more you know, the better you'll do'; the presenters of the show; an astrologer; two beauty bloggers who confessed to never having heard of any of the companies in question; and a cow named Gullros who would pick stocks by wandering around a field marked out in a grid of company names, and expressing her conviction by defecating in the relevant square.

The astrologer fared worst; the professionals did a little better, matching the performance of Gullros the cow (both the cow and the professionals achieved a respectable 7 per cent return over the three month contest); the beauty bloggers did better still – but the stand-out winners were the TV presenters, with a return of nearly 25 per cent over just three months. How had they done so well? Simple: they hadn't entered their own competition just once. Secretly, they'd done so twenty times by allowing themselves to pick twenty different portfolios. They revealed only the best-performing one to the audience. They appeared to be inspired stock-pickers, until they revealed their own trick. Survivor bias conquers all.[27]

With that in mind, it is hard to evaluate an investment manager who picks stocks or other financial products. They have everything to gain by persuading us that they are a genius, but have very little to show us except a track record. 'My fund beat the market last year, and the year before' is pretty much all we have to go on. The trouble is that we see only the successes, alongside the *schadenfreude* of the occasional high-profile implosion. Underperforming investment funds tend to be closed down, merged or rebranded. A major investment house will offer many different funds, and will advertise the ones that have been successful in the past. The Norwegian TV show condensed and exaggerated the process, but be assured that when fund managers advertise their stellar results, those adverts do not contain a random sample of the funds on offer.

Survivor bias even distorts some studies of investment performance. These studies often start by looking at 'funds that exist today' without fully acknowledging or adjusting for the fact that any fund still in existence is a survivor – and that introduces a survivorship bias. Burton Malkiel, economist and author of *A Random Walk Down Wall Street*, once tried to estimate how much survivorship bias flattered the performance of the surviving funds. His estimate – an astonishing 1.5 per cent per year. That might not sound like much, but over a lifetime of investing it's a factor of two: you expect retirement savings of (say) £100,000 and end up with £50,000 instead. Put another way, if you ignore all the investment funds that quietly disappear, the apparent performance is twice as good as the actual performance.[28] The result is to persuade people to invest in actively managed funds, which often charge high fees, when they might be better served by a low-cost, low-drama fund that passively tracks the stock market as a whole. That is a decision worth tens of billions of dollars a year across the US economy; if it's a mistake, it's a multi-billion-dollar mistake.[29]

So much for money. What about health? Consider the life-or-death matter of which medical treatments work and which don't. A randomised controlled trial (RCT) is often described as the 'gold standard' for medical evidence. In an RCT, some people receive the treatment being tested while others, chosen at random, are given either a placebo or the best known treatment. An RCT is indeed the fairest one-shot test of a new medical treatment, but if RCTs are subject to publication bias, we won't see the full picture of all the tests that have been done, and our conclusions are likely to be badly skewed.[30]

For example, in 2008 a quick survey of studies of a variety of antidepressant medications would have found forty-eight trials showing a positive effect, and three showing no positive effect. This sounds pretty encouraging, until you ponder the risk of publication bias. So the researchers behind that survey looked harder, digging out twenty-three unpublished trials, of which twenty-two had a negative result in which the drug did not help patients. They also found that eleven of the trials that seemed positive in the articles describing them had in fact produced negative results in the summaries presented to the regulator, the US Food and Drug Administration. The articles had managed to cherry-pick some good news and hand-wave away some bad news, and finish up presenting a positive-seeming picture about a drug that had not, in fact, been effective. The corrected score, then, was not 48–3 in favour of antidepressants working well, it was 38–37. Perhaps the antidepressants do work, at least sometimes or for some people, but it's fair to say that the published results did not fairly reflect all the experiments that had been conducted.[31]

This matters. Billions of dollars are misspent and hundreds of thousands of lives lost because of survivorship bias, when we make decisions without seeing the whole story – the investment funds that folded, the Silicon Valley entrepreneurs who never got beyond the 'junk in the garage' stage, the academic studies that were never published, and the clinical trials that went missing in action.

So far, this chapter has told a tale of catastrophe. The one bright spot is that these problems are vastly better understood and appreciated than they were even five years ago. So let's focus on that bright spot for a moment, and ask if there's hope for improvement.

For researchers, it's clear what that improvement would look like: they need to come clean about the Kickended side of research. They need to be transparent about the data that were gathered but not published, the statistical tests that were performed but then set to one side, the clinical trials that went missing in action, and the studies that produced humdrum results and were rejected by journals or stuffed in a file drawer while researchers got on with something more fruitful.

Those of us who write about research have a similar responsibility: not just to report on a stunning new result, but to set it in the context of what has been published before – and, preferably, what should have been published but languishes in obscurity.

Ideally, we need to be able to rise out of Andrew Gelman's 'garden of forking paths' and see the maze from above, including the dead ends and the paths less travelled. That view from above comes when we have all the relevant information in the most user-friendly form.

We are a long way from achieving those standards – but there are distinct signs of improvement. It is slow and incomplete, but it is improvement nonetheless. In medicine, for example, in 2005 the International Committee of Medical Journal Editors declared that the top medical journals they edited would no longer publish clinical trials that hadn't been preregistered. Preregistration means that before conducting a trial, researchers have to explain what they plan to do and how they plan to analyse the results, posting that explanation on a public website. Such preregistration is an important fix for publication bias, because it means that researchers can easily see cases in which a trial was planned but then somehow the results went missing in action. Preregistration should also allow other researchers to read a trial write-up and then go back to check that the plan for analysing the data was followed, rather than being changed once the data appeared.

Preregistration isn't a panacea. It poses a particular challenge for field studies in social science, which often require academic researchers to piggy-back on some project being conducted by a government or charitable organisation. Such projects evolve over time in ways that researchers cannot control or predict. And even when medical journals demand preregistration, they may fail to enforce their own demands.[32] Ben Goldacre and his colleagues at Oxford University's Centre for Evidence-Based Medicine spent a few weeks systematically monitoring the publication of new articles in the top medical journals. They identified fifty-eight articles that fell short of the reporting standards those journals had agreed to uphold – for example, clinical trials that had prespecified that they'd measure certain outcomes for patients, but then later switched to reporting different outcomes. They promptly wrote letters of correction to the journal editors but found that their letters were often rejected rather than published.[33]

It's disappointing to realise that standards are patchily enforced, but perhaps not surprising given that the entire system is basically self-regulated by the standards of a professional community, rather than governed by some central Solomonic figure. And it does seem to me that the situation has significantly improved over the past two decades: awareness is improving, bad practice is being called out, and it is better to have patchy standards than no standards at all. We have journals such as *Trials*, launched in 2006, which will publish the results of any clinical trial, regardless of whether the outcome was positive or negative, fascinating or dull, ensuring that no scientific study languishes unpublished simply because it wasn't regarded as newsworthy in the world of research. There's an enormous opportunity to do more with automated tools, such as automatically identifying missing trials, studies that were preregistered but then not published, or spotting when later papers are citing earlier research that has since been updated, corrected or withdrawn.[34]

In psychology, the kerfuffle over precognition may well have a positive result. Academic psychologists want to get published, of course, but most of them don't want to produce junk science; they want to find out what's true. The reproducibility crisis seems to be improving awareness of good research standards, as well as holding out more carrots to reward replication efforts, and more sticks to punish sloppy research.

There are encouraging signs that more researchers are welcoming replication efforts. For example, in 2010, political scientists Brendan Nyhan and Jason Reifler published a study on what became known as 'the backfire effect' – in brief, that people were more likely to believe a false claim if they'd been shown a fact-check that debunked the claim. This caused a moral panic among some journalists, particularly after the rise of Donald Trump. Fact-checking only makes matters worse! It hit that perfect counterintuitive sweet spot. But Nyhan and Reifler encouraged further studies, and those studies suggest that the backfire effect is unusual and fact-checking does help. One summary of the research concluded: 'generally debunking can make people's beliefs in specific claims more accurate'. Nyhan himself has quoted this summary on Twitter when he sees people relying on his original paper without considering the follow-ups.[35]

Many statisticians believe the crisis points to the need to rethink the standard statistical tests themselves – that the very concept of 'statistical significance' is deeply flawed. Mathematically, the test is simple enough. You start by assuming that there is no effect (the drug does not work; the coin is fair; precognition does not exist; the twenty-four-jam stall and the six-jam stall are equally appealing), and then you ask how unlikely the observed data are. For example, if you assume that a coin is fair and you toss it ten times, you'd expect to see heads five times, but you wouldn't be surprised to see six heads or maybe even seven. You'd be astonished to see ten heads in a row – and given that this would happen by chance less than one time in a thousand, you might question your original assumption that the coin was fair. Statistical significance testing relies on the same principle: assuming no effect, are the data you collect surprising? For instance, when testing a drug, your statistical analysis begins with the assumption that the drug does not work; when you observe that lots of the patients taking the drug are doing much better than the patients who are taking a placebo, you revise that assumption. In general, if the chances of randomly observing data at least as extreme as you collect are less than 5 per cent, the results are 'significant' enough to overturn the assumption: we can conclude with a sufficient degree of confidence that the drug works, large displays of jam discourage people from buying jam, and that precognition exists.

The problems are obvious. 5 per cent is an arbitrary cut-off point – why not 6 per cent, or 4 per cent? – and it encourages us to think in black-and-white, pass-or-fail terms, instead of embracing degrees of uncertainty. And if you found the previous paragraph confusing, I don't blame you. Conceptually, statistical significance is baffling, almost backwards: it tells us the chance of observing the data given a particular theory, the theory that there is no effect. Really, we'd like to know the opposite, the probability of a particular theory being true, given the data. My own instinct is that statistical significance is an unhelpful concept and we could do better, but others are more cautious. John Ioannidis – he of the 'Most Published Research Findings Are False' paper – argues that despite the flaws of the method, it's 'a convenient obstacle to unfounded claims'.

Unfortunately, there is no single clever statistical technique that would make all these problems evaporate. The journey towards more rigorous science requires many steps, and we at least are taking some of them. I recently had the chance to interview Richard Thaler, a Nobel Memorial Prize winner in economics, who has collaborated with Daniel Kahneman and many other psychologists. He struck me as well placed to evaluate psychology as a sympathetic outsider. 'I think the replication crisis has been great for psychology,' he told me. 'There's just better hygiene.'[36] Brian Nosek, meanwhile, told the BBC: 'I think if we do another large reproducibility project five years from now, we are going to see a dramatic improvement in reproducibility in the field.'[37]

In the early chapters of this book, I cited numerous psychological studies of motivated reasoning and the biased assimilation of information. You may by now be wondering: how do I know that those studies are credible?

The honest answer is that I cannot be certain. Any experimental research I cite has a chance of being the next jam experiment – or, much worse, the next discovery that listening to 'When I'm Sixty-Four' will make you younger. But when I read the studies I've described, I try to put the advice from the last few pages into practice. I try to get a sense of whether the study fits into the broader picture of what we know, or whether it's some strange outlier. If there are twenty or thirty studies from different academics using different methods, but all pointing to a similar conclusion – for instance that our powers of logical reasoning are skewed by our political beliefs – then I am less concerned that an individual experiment might turn out to be a fluke. If an empirical discovery makes sense in theory and in practice as well as in the lab, that's reassuring.

On most topics, most of us will not be digging through academic papers. We'll rely on the media to get a digestible take on the state of scientific knowledge. Science journalism is like any other kind of journalism: there is good, and there is bad. You can find superficial, sensationalist retreads of press releases that are themselves superficial and sensationalist. Or you can find science journalism that explains the facts, puts them in a proper context, and when necessary speaks truth to power. If you care enough as a reader you can probably figure out the difference. It's really not hard. Ask yourself if the journalist reporting on the research has clearly explained what's being measured. Was this a study done with humans? Or mice? Or in a petri dish? A good reporter will be clear. Then: how large is the effect? Was this a surprise to other researchers? A good journalist will try to make space to explain – and the article will be much more fun to read as a result, satisfying your curiosity and helping you to understand.*

If in doubt, you can easily find second opinions: almost any major research finding in science or social science will quickly be picked up and digested by academics and other specialists, who'll post their own thoughts and responses online. Science journalists themselves believe that the internet has improved their profession: in a survey of about a hundred European science journalists, two thirds agreed with that idea, and fewer than 10 per cent disagreed.[38] That makes sense: the internet has made it easier to read the journal articles, easier to access the systematic reviews, and easier to reach scientists for a second opinion.

If the story you're reading is about health, there's one place you should be sure to look for a second opinion: the Cochrane Collaboration. It's named after Archie Cochrane,

* Or try this. After reading an article or a Facebook post describing some cool finding, just ask yourself how you'd explain it to a friend. Do you know what the researchers did, and why, whether the research was a shock or exactly what experts would have expected? If your explanation is along the lines of, 'some boffins discovered that blueberries give you cancer', maybe you didn't read a good piece of journalism.

a doctor, epidemiologist and campaigner for better evidence in medicine. In 1941, when Cochrane was captured by the Germans and became a prisoner of war, he improvised a clinical trial. It was an astonishing combination of bravery, determination and humility. The prison camp was full of sick men – Cochrane was one of them – and he suspected that the illness was caused by a dietary deficiency, but he knew that he didn't know enough to confidently prescribe a treatment. Rather than slump into despair or follow a hunch, he managed to organise his fellow prisoners to test the effects of different diets, discovered what they were lacking, and provided incontrovertible evidence to the camp commandant. Vitamin supplements were duly procured, and many lives were saved as a result.[39]

In 1979, Cochrane wrote that 'it is surely a great criticism of our profession that we have not organised a critical summary, by specialty or subspecialty, adapted periodically, of all relevant randomised controlled trials'. After Cochrane's death, this challenge was taken up by Sir Iain Chalmers. In the early 1990s, Chalmers began assembling a collection of systematic reviews, at first just of the randomised trials conducted in the field of perinatal health – the care of pregnant women and their babies. The effort grew into an international network of researchers who review, rate, synthesise and publish the best available evidence on a huge variety of clinical topics.[40] They call themselves the Cochrane Collaboration and they maintain the Cochrane Library, an online database of systematic research reviews. The full database is not freely available in every country, but the accessible research summaries are, providing short descriptions of the state of knowledge based on randomised trials.

I looked at some recent research summaries, pretty much at random, to see what came up. One of the front-page summaries promised to evaluate 'Yoga for treating urinary incontinence in women'. Well, I don't practise yoga, don't suffer from urinary incontinence and am not a woman, so my evaluation of this report promised to be uncompromised by any actual knowledge about the topic.

Before I looked at what the Cochrane Library had to say, I typed 'Can Yoga Cure Incontinence?' into Google. WebMD was one of the top search results.[41] It reported that a new trial had shown dramatic improvements for older women, although it noted that the study was quite small. The *Daily Mail* picked up on the same study and reported it in a similar way: the improvements were big, but the study was small.[42] The top search result was from a private health care company:[43] it enthused about the spectacular results and did not mention how small the study was, although it did link through to the original research.[44]

None of this reporting is great, but neither is it terrible. To be honest, I expected worse. Nor is much harm likely to result. People may take up yoga with false hopes, or alternatively may take up yoga, get better, and then credit the yoga when in fact they would have got better anyway. But none of this would be disastrous.

Still, the media reports failed to give the back story. They simply regurgitated the scientific research without any indication of whether it accorded with, or contradicted, anything that had already been discovered.

The Cochrane Library, by contrast, aims to provide an accessible summary of everything we know about yoga and incontinence – if anything. It's also on the first page of Google search results. Cochrane is not a secret.

The Cochrane review, written in plain and unshowy language, is clear enough. There have only been two studies of the issue. Both of them were small. The evidence is weak, but what evidence there is suggests that for urinary incontinence yoga is better than nothing, and that mindfulness meditation is better than yoga. That's it – the result of a quick Google search and one minute scanning a page written in plain English. (Translations into many languages are available.) It would be nice, of course, if there was a vast and credible evidence base to lean on, but in this case, there isn't – and I'd rather know that. Thanks to the Cochrane summary we no longer have to guess if there's a pile of important evidence that we simply weren't told about.[45]

A related network, the Campbell Collaboration, aims to do the same thing for social policy questions in areas such as education and criminal justice. As these efforts gain momentum and resources, it will become easier for us to work out whether a study makes sense, and fits into a wider pattern of discoveries – or whether it's a $55,000 potato salad.

RULE SIX

Ask who is missing

> The power to not collect data is one of the most important and little-understood sources of power that governments have . . . By refusing to amass knowledge in the first place, decision-makers exert power over the rest of us.
>
> —ANNA POWELL-SMITH, MissingNumbers.org

Nearly seven decades ago, the noted psychologist Solomon Asch gave a simple task to 123 experimental subjects. They were shown two illustrations, one with three quite different lines and the other of a 'reference line', and Asch asked them to pick which of the three lines was the same length as the reference line. Asch had a trick up his sleeve: he surrounded each subject with stooges who would unanimously pick the wrong line. Confused, the experimental subjects were often (though not always) swayed by the errors of those around them.

The Asch experiments are endlessly fascinating, and I often find myself discussing them in my writing and talks: they are a great starting point for a conversation about the pressure we all feel to conform, and they provide a memorable window into human nature.

Or do they? The experiments are elegant and powerful, but like many psychologists Asch was working with material that came readily to hand: 1950s American college students. We shouldn't criticise him too much for that; Asch was simply harvesting the low-hanging fruit. It would have been troublesome for him to have collected a representative sample of all Americans, even harder to study an international sample, and impossible for him to have known what the study would have shown if it had been conducted not in 1952 but in 1972. (Others were to run the follow-up experiments, which found somewhat lower levels of conformity – perhaps a sign of student rebelliousness in the Vietnam era.)

It's all too tempting, however, to act as though Solomon Asch discovered an immutable and universal truth – to discuss the results of psychological experiments on a very specific type of person, in this case 1950s American students, as though they were experiments on the human race as a whole. I am guilty of this myself at times, especially when under the time pressure of a talk. But we should draw conclusions about human nature only after studying a broad range of people. Psychologists are increasingly acknowledging the problem of experiments that study only 'WEIRD' subjects – that is, Western, Educated and from Industrialised Rich Democracies.

By 1996, a Cochrane-style review of the literature found that Asch's experiment had inspired 133 follow-ups. The overall finding stood up, which is encouraging in the light of the previous chapter: conformity is a powerful and widespread effect, though it seemed to have weakened over time. But the obvious next question to ask is this: does conformity vary in its power depending on who is under pressure to conform to whom?

Disappointingly, the follow-up studies were not very diverse – most had been conducted in the United States and almost all with students – but the few exceptions were illuminating. For example, a 1967 experiment conducted with the Inuit of Baffin Island in Canada found lower levels of conformity than one conducted with the Temne people of Sierra Leone. I am no anthropologist, but reportedly the Inuit had a relaxed and individualistic culture, while the Temne society had strict social norms, at least at the time that these experiments were conducted. In general – and with some notable exceptions such as Japan – conformity in the Asch-inspired experiments has been lower in societies which sociologists viewed as individualist, and higher in societies viewed as collectivist, where social cohesion is more important.[1]

That implies Asch probably understated the power of conformity by studying subjects from America, an individualistic society. But then, accounts in both psychology textbooks and pop-science books often exaggerate how much conformity Asch found. (Asch's experimental subjects often rebelled against group pressure. Hardly any of them buckled every single time; much more commonly they tried to equivocate by varying their actions across repeated rounds of the experiment, sometimes agreeing with the group and sometimes staking out a lonely position.) By pure luck, these two biases in the popular understanding of Asch's findings may have effectively cancelled each other out.[2]

How much of the conformity pressure came because the group being studied was a monoculture? Would a more heterogeneous group leave more space for dissent? There are some tantalising hints of that possibility – for example, follow-up studies found that people conformed to groups of friends much more than they conformed to groups of strangers. And when Asch instructed his stooges to disagree with each other, conformity pressure evaporated: his subjects were happy to pick the correct choice, even if they were the only one doing so, as long as others were disagreeing among themselves. All this suggests that one cure for conformity is to make decisions with a diverse group of people, people who are likely to bring different ideas and assumptions to the table. But this practical tactic is hard to test as the original experiments and many of the follow-ups were on homogeneous groups. One can't help feeling that an opportunity has been missed.

It should, I think, make us feel uneasy that most accounts of Asch's results completely ignore the omission of people who might have acted differently, and whom he could easily have included. Solomon Asch taught in a proudly coeducational institution, Swarthmore College in Pennsylvania. Was it really necessary that not a single one of his experimental participants, neither the stooges nor the subjects, was female?

As it happens, follow-up studies suggest that all-male groups are less conformist than all-female groups. So, again, you could say it's a case of no harm, no foul: Asch might have seen even stronger evidence of conformity had he looked beyond young American males.[3] Still, gender does matter, and Asch could have studied its effects, or at least used mixed-gender groups. But it evidently didn't occur to him, and it's discomfiting how few subsequent reports on his experiment seem to care.

If Solomon Asch was the only researcher to have done this, we could wave it away as a historical curiosity. But Asch isn't alone; of course he isn't. His student Stanley Milgram conducted a notorious set of electric shock experiments at Yale University in the 1960s. Here's how I once described his experiments in the *Financial Times*:[4]

> [Milgram] recruited unsuspecting members of the public to participate in a 'study of memory'. On showing up at the laboratory, they drew lots with another participant to see who would be 'teacher' and who 'learner'. Once the learner was strapped into an electric chair, the teacher retreated into another room to take control of a shock machine. As the learner failed to answer questions correctly, the teacher was asked to administer steadily increasing electric shocks. Many proved willing to deliver possibly fatal shocks – despite having received a painful shock themselves as a demonstration, despite the learner having already complained of a heart condition, despite the screams of pain and the pleadings to be released from the other side of the wall, and despite the fact that the switches on the shock machine read 'Danger: Severe Shock, XXX'. Of course, there were no shocks – the man screaming from the nearby room was pretending. Yet the research exerts a horrifying fascination.

My article should have mentioned, if only in passing, that all forty of Milgram's experimental subjects were men. But I wasn't thinking about that particular issue at the time, and so – like many others before me – it didn't occur to me to check.

I hope it would now, because since writing that article I have interviewed Caroline Criado Perez about her book *Invisible Women*. Meeting her was fun – she strolled into the BBC with an adorable little dog who curled up in the corner of the studio and left us to talk about the gender data gap. Reading her book was less fun, because the incompetence and injustice she described was so depressing – from the makers of protective vests for police officers who forgot that some officers have breasts, to the coders of a 'comprehensive'

Apple health app who overlooked that some iPhone users menstruate.[5] Her book argues that all too often, the people responsible for the products and policies that shape our lives implicitly view the default customer – or citizen – as male. Women are an afterthought. Criado Perez argues that the statistics we gather are no exception to this rule: she makes abundantly clear how easy it is to assume that data reflect an impartial 'view from nowhere', when in fact they can contain deep and subtle biases.

Consider the historical under-representation of women in clinical trials. One grim landmark was thalidomide, which was widely taken by pregnant women to ease morning sickness only for it to emerge that the drug could cause severe disability and death to unborn children. Following this disaster, women of childbearing age were routinely excluded from trials, as a precaution. But the precaution only makes sense if one assumes that we can learn most of what we need to know by testing drugs only in men – a big assumption.[6]

The situation has improved, but many studies still do not disaggregate data to allow an exploration of whether there might be a different effect in men and in women. Sildenafil, for example, was originally intended as a treatment for angina. The clinical trial – conducted on men – revealed an unexpected side effect: magnificent erections. Now better known as Viagra, the drug hit the market as a treatment for erectile dysfunction. But sildenafil might have yet another unexpected benefit: it could be an effective treatment for period pain. We're not sure, as only one small and suggestive trial has yet been funded.[7] If the trial for angina had equally represented men and women, the potential to treat period pain might have been as obvious as the impact on erections.

This kind of sex-dependent effect is surprisingly common. One review of drug studies conducted in male and female rodents found that the drug being tested had a sex-dependent effect more than half of the time. For a long time, researchers into muscle-derived stem cells were baffled by why they sometimes regenerated and sometimes didn't. It seemed entirely arbitrary, until it occurred to someone to check whether the cells came from males or females. Mystery solved: it turned out that the cells from females regenerated, while those from males did not.

The gender blind-spot has yet to be banished. A few weeks into the coronavirus epidemic, researchers started to realise that men might be more susceptible than women, both to infection and to death. Was that because of a difference in behaviour, in diligent hand-washing, in the prevalence of smoking, or perhaps a deep difference in the biology of the male and female immune systems? It wasn't easy to say, particularly since of the twenty-five countries with the largest number of infections, more than half – including the UK and the US – did not disaggregate the cases by gender.[8]

A different problem arises when women are included in data-gathering exercises, but the questions they are asked don't fit the man-shaped box in the survey-designer's head. About twenty-five years ago in Uganda, the active labour force suddenly surged by over 10 per cent, from 6.5 to 7.2 million people. What had happened? The labour force survey started asking better questions.[9]

Previously, people had been invited to list their primary activity or job, and many women who held down part-time jobs, ran their own market stalls or put in hours on the family farm simply wrote down 'housewife'. The new survey asked about secondary activities as well, and suddenly women mentioned the long hours of paid work they had been doing on the side. Uganda's labour force increased by 700,000 people, most of them women. The problem was not that the women were ignored by the earlier survey, but that it asked questions that assumed an old-fashioned division of household labour in which the husband did full-time paid work and the wife worked unpaid in the home.

An even subtler gap in the data emerges from the fact that governments often measure the income not of individuals but of households. This is not an unreasonable decision: in a world where many families pool their resources in order to cover rent, food and sometimes all expenses, the 'household' is a logical unit of analysis. I know several people, men and women, who spend much or most of their time doing unpaid work at home, looking after children, while their partners are earning large salaries. It would be strange to claim that, on the basis that the unpaid partner earns little or no income, they live in poverty.

And yet while many households pool their resources, we cannot simply assume that they all do: money can be used as a weapon within a household, and unequal earnings can empower abusive relationships. Collecting data on household income alone makes such abuses statistically invisible, irrelevant by definition. It is all too tempting to assume that what we do not measure simply does not exist.

As with the Asch experiments, it turns out that we don't have to speculate that it might matter who within a household controls the purse strings. We have good evidence that it sometimes does. Economist Shelly Lundberg and colleagues studied what happened in the UK when in 1977, child benefit, a regular subsidy to families, was switched from being a tax credit (usually to the father) to a cash payment to the mother. That shift measurably increased spending on women's and children's clothes relative to men's clothes.[10]

When I wrote about Lundberg's research in the *Financial Times*, an outraged reader wrote to ask me how I knew that it was better to spend money on women's and children's clothes rather than men's clothes. Uncharacteristically for readers of the *FT*, this person had missed the point: it is not that any spending pattern was better, but that the spending pattern was different. Household income did not change, but when that income was paid to a different person in the household, it was spent on different things. That tells us that measuring income only at the level of the household omits important information. The UK's new benefit system, Universal Credit, is payable to a single 'head of household'. That curiously old-school decision may well favour men – but, given the data we have, it's going to be hard to tell.

It'd be nice to fondly imagine that high-quality statistics simply appear in a spreadsheet somewhere, divine providence from the numerical heavens. Yet any dataset begins with somebody deciding to collect the numbers. What numbers are and aren't collected, what is and isn't measured, and who is included or excluded, are the result of all-too-human assumptions, preconceptions and oversights.

The United Nations, for example, has embraced a series of ambitious 'Sustainable Development Goals' for 2030. But development experts are starting to call attention to a problem: we often don't have the data we would need to figure out whether those goals have been met. Are we succeeding in reducing the amount of domestic violence suffered by women? If few countries have chosen to collect good enough data on the problem to allow for historical comparisons, it's very hard to tell.[11]

Sometimes the choices about what data to gather are just bizarre. Will Moy, the director of the fact-checking organisation Full Fact, points out that in England, the authorities know more about golfers than they do about people who are assaulted, robbed or raped.[12] That's not because somebody in government sat down with a budget to commission surveys and decided that it was more important to understand golf than crime. Instead, surveys tend to be bundled with other projects. Amid the excitement of London being awarded the 2012 Olympic Games, the government launched the Active Lives Survey, which reaches 200,000 people, with enough geographical spread to allow us to understand which sports are most popular in each local area. That's why we know so much about golfers.

That's no bad thing – it's great to have such a fine-grained picture of how people keep fit. But doesn't it suggest there's a case for beefing up the Crime Survey of England and Wales, which reaches just 35,000 households? That's a large enough survey to understand the national trend in common crimes, but if it were as large as the Active Lives Survey, we might be able to understand trends for rare crimes, smaller demographic groups or particular towns. Other things being equal, a larger survey can give more precise estimates, especially when you're trying to count something unusual.

But bigger isn't always better. It's perfectly possible to reach vast numbers of people while still missing out enough other people to get a disastrously skewed impression of what's really going on.

In 1936, the Kansas Governor Alfred Landon was the Republican nominee for President against the incumbent, President Franklin Delano Roosevelt, a Democrat. A respected magazine, the *Literary Digest*, shouldered the responsibility of forecasting the result. It conducted an astonishingly ambitious postal opinion poll, which reached 10 million people, a quarter of the electorate. The deluge of mailed-in replies can hardly be imagined, but the *Digest* seemed to be relishing the scale of the task. In late August it reported, 'Next week, the first answers from these ten million will begin the incoming tide of marked ballots, to be triple-checked, verified, five-times cross-classified and totalled.'[13]

After tabulating a remarkable 2.4 million returns as they flowed in over two months, the *Literary Digest* announced its conclusions: Landon would win by a convincing 55 per cent to 41 per cent, with a few voters favouring a third candidate.

The election delivered a very different result. Roosevelt crushed Landon by 61 per cent to 37 per cent. To add to the *Literary Digest*'s agony, a far smaller survey conducted by the opinion poll pioneer George Gallup came much closer to the final vote, forecasting a comfortable victory for Roosevelt.

Mr Gallup understood something that the *Literary Digest* did not: when it comes to data, size isn't everything. Opinion polls such as Gallup's are based on samples of the voting population. This means opinion pollsters need to deal with two issues: sample error and sample bias.

Sample error reflects the risk that, purely by chance, a randomly chosen sample of opinions does not reflect the true views of the population. The 'margin of error' reported in opinion polls reflects this risk, and the larger the sample, the smaller the margin of error. A thousand interviews is a large enough sample for many purposes, and during the 1936 election campaign Mr Gallup is reported to have conducted three thousand interviews.

But if three thousand interviews were good, why weren't 2.4 million far better? The answer is that sampling error has a far more dangerous friend: sampling bias. Sampling error is when a randomly chosen sample doesn't reflect the underlying population purely by chance; sampling bias is when the sample isn't randomly chosen at all. George Gallup took pains to find an unbiased sample because he knew that was far more important than finding a big one.

Literary Digest, in its quest for a bigger dataset, fumbled the question of a biased sample. It mailed out forms to people on a list it had compiled from automobile registrations and telephone directories – a sample that, at least in 1936, was disproportionately prosperous. Those who had telephones or cars were generally wealthier than those who did not. To compound the problem, Landon supporters turned out to be more likely to mail back their answers than those who backed Roosevelt. The combination of those two biases was enough to doom the *Literary Digest*'s poll. For each person George Gallup's pollsters interviewed, *Literary Digest* received eight hundred responses. All that gave the magazine for its pains was a very precise estimate of the wrong answer. By failing to pay enough attention both to the missing people (the ones who were never surveyed) and to the missing responses, the *Literary Digest* perpetrated one of the most famous polling disasters in statistical history.

All pollsters know that their polls are vulnerable to the *Literary Digest* effect, and the serious ones try – as George Gallup tried – to reach a representative sample of the population. This has never been easy, and it seems to be getting harder: fewer people bother to answer the pollsters' enquiries, raising the obvious question about whether the ones who do are really representative of everyone else. This is partly because people are less willing to pick up a landline telephone to speak to cold-callers, but that's not the only explanation. For example, the first British Election Study, a face-to-face survey in which the survey team would knock on people's doors, had a response rate of nearly 80 per cent back in 1963. The 2015 version, also face-to-face, had a response rate of just over 55 per cent; in almost half of the homes approached, either nobody opened the door, or somebody opened the door but refused to answer the surveyor's questions.[14]

Pollsters try to correct for this, but there is no foolproof method of doing so. The missing responses are examples of what the statistician David Hand calls 'dark data': we know the people are out there and we know that they have opinions, but we can only guess at what those opinions are. We can ignore dark data, as Asch and Milgram ignored the question of how women would respond in their experiments, or we can try desperately to shine a light on what's missing. But we can never entirely solve the problem.

In the UK General Election of 2015, opinion polls suggested that David Cameron, the incumbent Prime Minister, was unlikely to win enough votes to stay in power. The polls were wrong: Cameron's Conservative party actually gained seats in the House of Commons and secured a narrow victory. It was unclear what had gone wrong, but many polling companies presumed that there had been a last-minute swing in favour of the Conservatives. If only they had conducted a few snap polls at the last possible moment, they might have detected that swing.

But that diagnosis of what had gone wrong was incorrect. Later research showed that the real problem was dark data. Shortly after the election, researchers chose a random sample of houses and knocked on the door to ask people if and how they voted. They got the same answer as the pollsters had: not enough Conservative voters to return Mr Cameron to office. But the pollsters then went back again to the houses where nobody had answered, or where people had turned the surveyors away. On the second attempt, more Conservative voters were in evidence. The pollsters came back to try to fill in the gaps again, and again, and again – sometimes as many as six times – and eventually got an answer from almost everyone they'd originally hoped to talk to. The conclusion: this retrospective poll finally matched the result of the election – a Conservative government.

If the problem had been a late swing, the solution would have been a bunch of quick-and-dirty, last-minute surveys. But because the real problem was that Conservative voters were harder to reach, the real solution may have to be a slower, more exhaustive method of conducting opinion polls.[15]

Both problems hit US pollsters in the notorious 2016 election, when the polls seemed to put Hillary Clinton ahead of Donald Trump in the swing states that would decide the contest. There was a late swing towards Trump, and also the same kind of non-response bias that had doomed the 2015 UK polls: it turned out to have been easier for pollsters to find Clinton supporters than Trump supporters. The polling error was not, objectively speaking, very large. It just loomed large in people's imagination, perhaps because Trump was such an unusual candidate. But the fact remains that the polls were wrong in part because when the pollsters tried to find a representative group of voters to talk to, too many Trump supporters were missing.[16]

One ambitious solution to the problem of sample bias is to stop trying to sample a representative slice of the population, and instead speak to everybody. That is what the census attempts to do. However, even census-takers can't assume they have counted everyone. In the US 2010 census, they received responses from just 74 per cent of households. That's a lot of people missing out or opting out.

In the UK 2011 census, the response rate was 95 per cent, representing about 25 million households. That's much better – indeed, it seems almost perfect at first glance. With

25 million households responding, random sample error is not an issue; it will be tiny. But even with just 5 per cent of people missing, sample bias is still a concern. The census-takers know that certain kinds of people are less likely to respond when the official-looking census form lands with a thud on the doormat: people who live in multiple-occupancy houses such as a shared student house; men in their twenties; people who don't speak good English. As a result, the 5 per cent who don't respond may look very different from the 95 per cent who do. That fact alone is enough to skew the census data.[17]

Census-taking is among the oldest ways of collecting statistics. Much newer, but with similar aspirations to reach everyone, is 'big data'. Professor Viktor Mayer-Schönberger of Oxford's Internet Institute, and co-author of the book *Big Data*, told me that his favoured definition of a big dataset is one where 'N = All' – where we no longer have to sample, because we have the entire background population.[18]

One source of big data is so mundane as to be easy to overlook. Think about the data you create when you watch a movie. In 1980 your only option would have been to go to a cinema, probably paying with cash. The only data created would have been box office receipts. In 1990 you could instead have gone to your local video rental store; they might have had a computer to track your rental, or it might have all been done with pen and paper. If it was done on a computer it would probably not have been connected to any broader database. But in the twenty-first century, when you sign up for an account with Netflix or Amazon, your data enter a vast and interconnected world – easily analysed, cross-referenced or shared with a data wholesaler, if the terms and conditions allow.

The same story is true when you apply for a library card,

pay income tax, sign up for a mobile phone contract or apply for a passport. Once upon a time, such data would have existed as little slips of paper in a giant alphabetical catalogue. They weren't designed for statistical analysis, as a census or a survey would have been. They were administrative building blocks – data gathered in order to get things done. Over time, as administrative data have been digitised and algorithms that can interrogate the data have been improved, it has become ever easier to use them as an input to statistical analysis, a complement to or even a substitute for survey data.

But 'N = All' is often more of a comforting assumption than a fact. As we've seen, administrative data will often include information about whoever in the household fills in the forms and pays the bills; the admin-shy will be harder to pin down. And it is all too easy to forget that 'N = All' is not the same as 'N = Everyone who has signed up for a particular service'. Netflix, for example, has copious data about every single Netflix customer, but far less data about people who are not Netflix customers – and it would be perilous for Netflix to generalise from one group to the other.

Even more than administrative data, the lifeblood of big data is 'found data' – the kind of data we leave in our wake without even noticing, as we carry our smartphones around, search Google, pay online, tweet our thoughts, post photos to Facebook, or crank up the heating on our smart thermostat. It's not just the name and credit card details that you gave to Netflix: it's everything you ever watched on the streaming service, when you watched it – or stopped watching it – and much else besides.

When data like these are opportunistically scraped from cyberspace, they may be skewed in all sorts of awkward ways. If we want to put our finger on the pulse of public opinion, for example, we might run a sentiment analysis algorithm on Twitter rather than going to the expense of commissioning an opinion poll. Twitter can supply every message for analysis, although in practice most researchers use a subset of that vast firehose of data. But even if we analysed every Twitter message – N = All – we would still learn only what Twitter users think, not what the wider world thinks. And Twitter users are not particularly representative of the wider world. In the United States, for example, they are more likely than the population as a whole to be young, urban, college-educated and black. Women, meanwhile, are more likely than men to use Facebook and Instagram, but less likely to use LinkedIn. Hispanics are more likely than whites to use Facebook, while blacks are more likely than whites to use LinkedIn, Twitter and Instagram. None of these facts is obvious.[19]

Kate Crawford, a researcher at Microsoft, has assembled many examples of when N = All assumptions have led people astray. When Hurricane Sandy hit the New York area in 2012, researchers published an analysis of data from Twitter and a location-based search engine, FourSquare, showing that they could track a spike in grocery shopping the day before and a boom for bars and nightclubs the day after. That's fine, as far as it goes – but those tweets about the hurricane were disproportionately from Manhattan, whereas areas such as Coney Island had been hit much harder. In fact, Coney Island had been hit so hard the electricity was out – that was why nobody there was tweeting – while densely populated and prosperous Manhattan was unusually saturated with smartphones, at least by 2012 standards, when they were less ubiquitous than today. To make this sort of big data analysis useful, it takes a considerable effort to disentangle the tweets from the reality.[20]

Another example: in 2012 Boston launched a smartphone

app, StreetBump, which used an iPhone's accelerometer to detect potholes. The idea was that citizens of Boston would download the app and, as they drove around the city, their phones would automatically notify City Hall when the road surface needed repair – city workers would no longer have to patrol the streets looking for potholes. It's a pleasingly elegant idea, and it did successfully find some holes in the road. Yet what StreetBump really produced, left to its own devices, was a map of potholes that systematically favoured young, affluent areas where more people owned iPhones and had heard about the app. StreetBump offers us 'N = All' in the sense that every bump from every enabled phone can be recorded. That is not the same thing as recording every pothole. The project has since been shelved.

The algorithms that analyse big data are trained using found data that can be subtly biased. Algorithms trained largely on pale faces and male voices, for example, may be confused when they later try to interpret the speech of women or the appearance of darker complexions. This is believed to help explain why Google photo software confused photographs of people with dark skin with photographs of gorillas; Hewlett Packard webcams struggled to activate when pointing at people with dark skin tones; and Nikon cameras, programmed to retake photographs if they thought someone had blinked during the shot, kept retaking shots of people from China, Japan or Korea, mistaking the distinctively east Asian eyelid fold for a blink. New apps, launched in the spring of 2020, promise to listen to you cough and detect whether you have Covid-19 or some other illness. I wonder whether they will do better?[21]

One thing is certain. If algorithms are shown a skewed sample of the world, they will reach a skewed conclusion.[22]

*

There are some overtly racist and sexist people out there – look around – but in general what we count and what we fail to count is often the result of an unexamined choice, of subtle biases and hidden assumptions that we haven't realised are leading us astray.

Unless we're collecting data ourselves, there's a limit to how much we can do to combat the problem of missing data. But we can and should remember to ask who or what might be missing from the data we're being told about. Some missing numbers are obvious – for example, it's clearly hard to collect good data about crimes such as sex-trafficking or the use of hard drugs. Other omissions show up only when we take a close look at the claim in question. Researchers may not be explicit that an experiment only studied men – such information is sometimes buried in a statistical appendix, and sometimes not reported at all. But often a quick investigation will reveal that the study has a blind spot. If an experiment studies only men, we can't assume it would have pointed to the same conclusions if it had also included women. If a government statistic measures the income *of* a household, we must recognise that we're learning little about the sharing of that income *within* a household.

Big found datasets can seem comprehensive, and may be enormously useful, but 'N = All' is often a seductive illusion: it's easy to make unwarranted assumptions that we have everything that matters. We must always ask who and what is missing. And this is only one reason to approach big data with caution. Big data represents a huge and under-scrutinised change in the way statistics are being collected, and that is where our journey to make the world add up will take us next.

RULE SEVEN

Demand transparency when the computer says 'no'

> I know I've made some very poor decisions recently but I can give you my complete assurance that my work will be back to normal. I've still got the greatest enthusiasm and confidence in the mission. And I want to help you, Dave.
>
> —HAL 9000 *(2001: A Space Odyssey)*

In 2009, a team of researchers from Google announced a remarkable achievement in one of the world's top scientific journals, *Nature*.[1] Without needing the results of a single medical check-up, they were able to track the spread of influenza across the United States. What's more, they could do it more quickly than the Centers for Disease Control and Prevention (CDC), which relied on reports from doctors' surgeries. Google's algorithm had searched for patterns in the CDC data from 2003 to 2008, identifying a correlation between flu cases and what people had been searching for online in the same area at the same time. Having discovered the pattern, the algorithm could now use today's searches to estimate today's flu cases, a week or more before the CDC published its official view.[2]

Not only was 'Google Flu Trends' quick, accurate and cheap, it was theory-free. Google's engineers didn't bother to develop a hypothesis about what search terms might be correlated with the spread of the disease. We can reasonably guess that searches such as 'flu symptoms' or 'pharmacies near me' might better predict flu cases than searches for 'Beyoncé', but the Google team didn't care about that. They just fed in their top 50 million search terms and let the algorithms do the work.

The success of Google Flu Trends became emblematic of the hot new trend in business, technology and science: 'big data' and 'algorithms'. 'Big data' can mean many things, but let's focus on the found data we discussed in the previous chapter, the digital exhaust of web searches, credit card payments and mobiles pinging the nearest cellphone mast, perhaps buttressed by the administrative data generated as organisations organise themselves.

An algorithm, meanwhile, is a step-by-step recipe* for performing a series of actions, and in most cases 'algorithm' means simply 'computer program'. But over the past few years, the word has come to be associated with something quite specific: algorithms have become tools for finding patterns in large sets of data. Google Flu Trends was built on pattern-recognising algorithms churning through those 50 million search terms, looking for ones that seemed to coincide with the CDC reporting more cases of flu.

It's these sorts of data, and these sorts of algorithms, that I'd like to examine in this chapter. 'Found' datasets can be

---

* Although if it is a recipe, it is a recipe written by a particularly pedantic chef. Most recipes leave room for common sense, but if an algorithm is to be interpreted by a computer the steps must be tightly specified.

huge. They are also often relatively cheap to collect, updated in real time, and messy – a collage of data points collected for disparate purposes. As our communication, leisure and commerce are moving to the internet, and the internet is moving into our phones, our cars and even our spectacles, life can be recorded and quantified in a way that would have been hard to imagine just a decade ago. The business bookshelves and the pages of management magazines are bulging with books and articles on the opportunities such data provide.

Alongside a 'wise up and get rich' message, cheerleaders for big data have made three exciting claims, each one reflected in the success of Google Flu Trends. First, that data analysis produces uncannily accurate results. Second, that every single data point can be captured – the 'N = All' claim we met in the last chapter – making old statistical sampling techniques obsolete (what that means here is that Flu Trends captured every single search). And finally, that scientific models are obsolete, too: there's simply no need to develop and test theories about why searches for 'flu symptoms' or 'Beyoncé' might or might not be correlated with the spread of flu, because, to quote a provocative *Wired* article in 2008, 'with enough data, the numbers speak for themselves'.

This is revolutionary stuff. Yet four years after the original *Nature* paper was published, *Nature News* had sad tidings to convey: the latest flu outbreak had claimed an unexpected victim – Google Flu Trends. After reliably providing a swift and accurate account of flu outbreaks for several winters, the theory-free, data-rich model lost its nose for where flu was going. Google's model pointed to a severe outbreak, but when the slow-and-steady data from the CDC arrived, they showed that Google's estimates of the spread of flu-like illnesses were overstated – at one point they were more than double the true figure.[3] The Google Flu Trends project was shut down not long after.[4]

What had gone wrong? Part of the problem was rooted in that third exciting claim: Google did not know – and it could not begin to know – what linked the search terms with the spread of flu. Google's engineers weren't trying to figure out what caused what. They were merely finding statistical patterns in the data, which is what these algorithms do. In fact, the Google research team had peeked into the patterns and discovered some obviously spurious correlations that they could safely instruct the algorithm to disregard – for example, flu cases turned out to be correlated with searches for 'high school basketball'. There's no mystery about why: both flu and high school basketball tend to get going in the middle of November. But it meant that Flu Trends was part flu detector, part winter detector.[5] That became a problem when there was an outbreak of summer flu in 2009: Google Flu Trends, eagerly scanning for signs of winter and finding nothing, missed the non-seasonal outbreak, as true cases were four times higher than it was estimating.[6]

The 'winter detector' problem is common in big data analysis. A literal example, via computer scientist Sameer Singh, is the pattern-recognising algorithm that was shown many photos of wolves in the wild, and many photos of pet husky dogs. The algorithm seemed to be really good at distinguishing the two rather similar canines; it turned out that it was simply labelling any picture with snow as containing a wolf. An example with more serious implications was described by Janelle Shane in her book *You Look Like a Thing and I Love You*: an algorithm that was shown pictures of healthy skin and of skin cancer. The algorithm figured out the pattern: if there was a ruler in the photograph, it was cancer.[7] If we don't know *why* the algorithm is doing what it's doing, we're trusting our lives to a ruler detector.

Figuring out what causes what is hard – impossible, some

say. Figuring out what is correlated with what is much cheaper and easier. And some big data enthusiasts – such as Chris Anderson, author of that provocative article in *Wired* magazine – have argued that it is pointless to look beyond correlations. 'View data mathematically first and establish a context for it later,' he wrote; the numbers speak for themselves. Or, to rephrase Anderson's point unkindly, 'If searches for high school basketball always pick up at the same time as flu cases, it doesn't really matter why'.

But it *does* matter, because a theory-free analysis of mere correlations is inevitably fragile. If you have no idea what is behind a correlation, you have no idea what might cause that correlation to break down.

After the summer flu problem of 2009, the accuracy of Flu Trends collapsed completely at the end of 2012. It's not clear why. One theory is that the news was full of scary stories about flu in December 2012, and these stories might have provoked internet searches from people who were healthy. Another possible explanation is a change in Google's own search algorithm: it began automatically suggesting diagnoses when people entered medical symptoms, and this will have changed what they typed into Google in a way that might have foxed the Flu Trends model. It's quite possible that Google could have figured out what the problem was and found a way to make the algorithm work again if they'd wanted to, but they just decided that it wasn't worth the trouble, expense and risk of failure.

Or maybe not. The truth is, external researchers have been forced to guess at exactly what went wrong, because they don't have the information to know for sure. Google shares some data with researchers, and indeed makes some data freely available to anyone. But it isn't going to release all its data to you, or me, or anyone else.

*

Two good books with pride-of-place on my bookshelf tell the story of how our view of big data evolved over just a few short years.

One, published in 2013, is *Big Data* by Kenn Cukier and Viktor Mayer-Schönberger. It reports many examples of how cheap sensors, huge datasets and pattern-recognising algorithms were, to paraphrase the book's subtitle, 'transforming how we live, work and think'. The triumphant example the authors chose to begin their story? Google Flu Trends. The collapse became apparent only after the book had gone to print.

Three years later, in 2016, came Cathy O'Neil's *Weapons of Math Destruction*, which – as you might be able to guess – takes a far more pessimistic view. O'Neil's subtitle tells us that big data 'increases inequality and threatens democracy'.

The difference is partly one of perspective: Cukier and Mayer-Schönberger tend to adopt the viewpoint of someone who is doing something with a data-driven algorithm; O'Neil tends to see things from the viewpoint of someone to whom a data-driven algorithm is doing something. A hammer looks like a useful tool to a carpenter; the nail has a different impression altogether.

But the change in tone also reflects a change in the zeitgeist between 2013 and 2016. In 2013, the relatively few people who were paying attention to big data often imagined themselves to be the carpenters; by 2016, many of us had realised that we were nails. Big data went from seeming transformative to seeming sinister. Cheerleading gave way to doomsaying, and some breathlessly over-the-top headlines. (Perhaps my favourite was a CNN story: 'Math is Racist'.) The crisis reached a shrill pitch with the discovery that a political consulting firm, Cambridge Analytica, had exploited Facebook's lax policies on data to slurp up information about 50 million people, without those people knowing or meaningfully consenting to this, and show them personally targeted advertisements. It was briefly supposed by horrified commentators that these ads were so effective they essentially elected Donald Trump, though more sober analysis later concluded that Cambridge Analytica's capabilities fell short of mind control.[8]

Each of us is sweating data, and those data are being mopped up and wrung out into oceans of information. Algorithms and large datasets are being used for everything from finding us love to deciding whether, if we are accused of a crime, we go to prison before the trial or are instead allowed to post bail. We all need to understand what these data are and how they can be exploited. Should big data excite or terrify us? Should we be more inclined to cheer the carpenters or worry about our unwitting role as nails?

The answer is that it depends – and in this chapter I hope to show you what it depends on.

Writing in the *New York Times* magazine in 2012, when the zeitgeist was still firmly with the carpenters, the journalist Charles Duhigg captured the hype about big data quite brilliantly with an anecdote about the US discount department store Target.

Duhigg explained that Target had collected so much data on its customers, and was so skilled at analysing those data, that its insight into consumers could seem like magic.[9] His killer anecdote was of the man who stormed into a Target near Minneapolis and complained to the manager that the company was sending coupons for baby clothes and maternity wear to his teenage daughter. The manager apologised profusely, and later called to apologise again – only to be told that the teenager was indeed pregnant. Her father hadn't realised.

Target, after analysing her purchases of unscented wipes and vitamin supplements, had.

Yet is this truly statistical sorcery? I've discussed this story with many people and I'm struck by a disparity. Most people are wide-eyed with astonishment. But two of the groups I hang out with a lot take a rather different view. Journalists are often cynical; some suspect Duhigg of inventing, exaggerating or passing on an urban myth. (I suspect *them* of professional jealousy.) Data scientists and statisticians, on the other hand, yawn. They regard the tale as both unsurprising and uninformative. And I think the statisticians have it right.

First, let's think for a moment about just how amazing it might be to predict that someone is pregnant based on their shopping habits: not very. Consider the National Health Service advice on the vitamin supplement folic acid:

> It's recommended that all women who could get pregnant should take a daily supplement of 400 micrograms of folic acid before they're pregnant and during the first 12 weeks of pregnancy . . . If you did not take folic acid supplements before getting pregnant, you should start taking them as soon as you find out you're pregnant . . . The only way to be sure you're getting the right amount is by taking a supplement.

OK. With this in mind, what conclusion should I reach if I am told that a woman has started buying folic acid? I don't need to use a vast dataset or a brilliant analytical process. It's not magic. Clearly, she might well be pregnant. The Target algorithm hadn't produced a superhuman leap of logic, but a very human one: it figured out exactly what you or I or anyone else would also have figured out, given the same information.

Admittedly, sometimes we humans can be slow on the uptake. Hannah Fry, the author of another excellent book about algorithms, *Hello World*, relays the example of a woman shopping online at the UK supermarket Tesco.[10] She discovered condoms in the 'buy this again?' section of her online shopping cart – implying that the algorithm knew someone in her household had bought them before. But she hadn't, and her husband had no reason to: they didn't use condoms together. So she assumed it was a technical error. I mean, what other explanation could there possibly be?

When the woman contacted Tesco to complain, the company representatives concluded that it wasn't their job to break the bad news that her husband was cheating on her, and went for the tactful white lie. 'Indeed, madam? A computer error? You're quite right, that must be the reason. We are so sorry for the inconvenience.' Fry tells me that this is now the rule of thumb at Tesco: apologise, and blame the computer.

If a customer previously bought condoms, they might want to buy them again. If someone bought a pregnancy test and then starts buying a vitamin supplement designed for pregnant women, it's a reasonable bet that this person is a woman and that in a few months they might become interested in buying maternity wear and baby clothes. The algorithms aren't working statistical miracles here. They're simply seeing something (the condoms, the pregnancy vitamins) that has been concealed from the human (the puzzled wife, the angry dad). We're awestruck by the algorithm in part because we don't appreciate the mundanity of what's happening underneath the magician's silk handkerchief.

And there's another way in which Duhigg's story of Target's algorithm invites us to overestimate the capabilities of data-fuelled computer analytics.

'There's a huge false positive issue,' says Kaiser Fung, a data scientist who has spent years developing similar approaches for retailers and advertisers. What Mr Fung means is that we don't get to hear stories about women who receive coupons for baby wear but who aren't pregnant. Hearing the anecdote, it's easy to assume that Target's algorithms are infallible – that everybody receiving coupons for onesies and wet wipes is pregnant. But nobody ever claimed that it was true. And it almost certainly isn't; maybe *everybody* received coupons for onesies. We should not simply accept the idea that Target's computer is a mind-reader before considering how many misses attend each hit.

There may be many of those misses – even for an easy guess such as 'woman buying folic acid may be pregnant'. Folic acid purchases don't guarantee pregnancy. The woman might be taking folic acid for some other reason. Or she might be buying the vitamin supplement for someone else. Or – and imagine the distress when the vouchers for baby clothes start arriving – she might have been pregnant but miscarried, or trying to get pregnant without success. It's possible that Target's algorithm is so brilliant that it filters out these painful cases. But it is not likely.

In Charles Duhigg's account, Target mixes in random offers, such as coupons for wine glasses, because pregnant customers would feel spooked if they realised how intimately the company's computers understood them. But Kaiser Fung has another explanation: Target mixes up its offers not because it would be weird to send an all-baby coupon book to a woman who was pregnant, but because the company knows that many of those coupon books will be sent to women who aren't pregnant after all.

The manager should simply have said that: don't worry about it, lots of people get these coupons. Why didn't he? He probably didn't understand the Target algorithm any more than the rest of us. Just like Google, Target might be reluctant to open up its algorithm and dataset for researchers – and competitors – to understand what's going on.

The most likely situation is this: pregnancy is often a pretty easy condition to spot through shopping behaviour, so Target's big data-driven algorithm surely can predict pregnancy better than random guessing. Doubtless, however, it is well short of infallibility. A random guess might be that any woman between fifteen and forty-five has a roughly 5 per cent chance of being pregnant at any time. If Target can improve to guessing correctly 10 or 15 per cent of the time, that's already likely to be well worth doing. Even a modest increase in the accuracy of targeted special offers would help the bottom line. But profitability should not be conflated with omniscience.

So let's start by toning down the hype a little – both the apocalyptic idea that Cambridge Analytica can read your mind, and the giddy prospect that big data can easily replace more plodding statistical processes such as the CDC's survey of influenza cases. When I first started grappling with big data, I called Cambridge University's Professor Sir David Spiegelhalter – one of the country's leading statisticians, and a brilliant statistical communicator. I summarised the cheerleading claims: of uncanny accuracy, of making sampling irrelevant because every data point was captured, and of consigning scientific models to the junk-heap because 'the numbers speak for themselves'.

He didn't feel the need to reach for a technical term. Those claims, he said, are 'complete bollocks. Absolute nonsense.'

Making big data work is harder than it seems. Statisticians have spent the past two hundred years figuring out what traps lie in wait when we try to understand the world through data. The data are bigger, faster and cheaper these days, but we must not pretend that the traps have all been made safe. They have not.

'There are a lot of small data problems that occur in big data,' added Spiegelhalter. 'They don't disappear because you've got lots of the stuff. They get worse.'

It hardly matters if some of Charles Duhigg's readers are too credulous about the precision with which Target targets onesie coupons. But it does matter when people in power are similarly overawed by algorithms they don't understand, and use them to make life-changing decisions.

One of Cathy O'Neil's most powerful examples in *Weapons of Math Destruction* is the IMPACT algorithm used to assess teachers in Washington DC. As O'Neil describes, much-loved and highly respected teachers in the city's schools were being abruptly sacked after achieving very poor ratings from the algorithm.

The IMPACT algorithm claimed to measure the quality of teaching, basically by checking whether the children in a teacher's class had made progress or slipped backwards in their test scores.[11] Measuring the true quality of teaching, however, is hard to do for two reasons. The first is that no matter how good or how bad the teacher, individual student achievement will vary a lot. With just thirty students in a class, a lot of what the algorithm measures will be noise; if a couple of kids made some lucky guesses in their start-of-year test and then were unlucky at the end of year, that's enough to make a difference to a teacher's ranking. It shouldn't be, because it's pure chance. Another source of variation that's out of the teacher's control is when a child has a serious problem outside the classroom – anything from illness to bullying to a family member being imprisoned. This isn't the same kind of noise as lucky or unlucky guesses in a test, because it tracks something real. A system that tracked and followed up signs of trouble outside class would be valuable. But it would be foolish and unfair to blame the pupil's struggles on the teacher.

The second problem is that the algorithm may be fooled by teachers who cheat, and the cheats will damage the prospects of the honest teachers. If the sixth-grade teacher finds a way to unfairly boost the test performance of the children he teaches – such things aren't unknown – then not only will he be unfairly rewarded, but the seventh-grade teacher is going to be in serious trouble next year. Her incoming class will be geniuses on paper; improvement will be impossible unless she also finds a way to cheat.

O'Neil's view, which is plausible, is that the data are so noisy that the task of assessing a teacher's competence is hopeless for any algorithm. Certainly this particular algorithm's judgements on which teachers were sub-par didn't always tally with those of the teachers' colleagues or students. But that didn't stop the DC school district authorities firing 206 teachers in 2011 for failing to meet the algorithm's standards.

So far we've focused on excessive credulity in the power of the algorithm to extract wisdom from the data it is fed. There's another, related problem: excessive credulity in the quality or completeness of the dataset.

We explored the completeness problem in the previous chapter. *Literary Digest* accumulated what might fairly be described as big data. It was certainly an enormous survey by the standards of the day – indeed, even by today's standards a dataset with 2.4 million people in it is impressive. But you can't use *Literary Digest* surveys to predict election results if 'people who respond to *Literary Digest* surveys' differ in some consistent way from 'people who vote in elections'.

Google Flu Trends captured every Google search, but not everybody who gets flu turns to Google. Its accuracy depended on 'people with flu who consult Google about it' not being systematically different from 'people with flu'. The pothole-detecting app we met in the last chapter fell short because it confused 'people who hear about and install pothole-detecting apps' with 'people who drive around the city'.

How about quality? Here's an instructive example of big data from an even older vintage than the 1936 US election poll: the astonishing attempt to assess the typical temperature of the human body. Over the course of eighteen years, the nineteenth-century German doctor Carl Wunderlich assembled over a million measurements of body temperature, gathered from more than 25,000 patients. A million measurements! It's a truly staggering achievement given the pen-and-paper technology of the day. Wunderlich is the man behind the conventional wisdom that normal body temperature is 98.6°F. Nobody wanted to gainsay his findings, partly because the dataset was large enough to command respect, and partly because the prospect of challenging it with a bigger, better dataset was intimidating. Dr Philip Mackowiak, an expert on Wunderlich, put it, 'Nobody was in a position or had the desire to amass a dataset that large.'[12]

Yet Wunderlich's numbers were off; we're normally a little cooler (by about half a Fahrenheit degree).[13] So formidable were his data that it took more than a hundred years to establish that the good doctor had been in error.*

* The problem was exacerbated by a conversion of units. Wunderlich's original measurements were made in centigrade, and his results concluded that the typical body temperature was a range around 37°C – implicitly, given that degree of precision, a range of up to a degree centigrade, somewhere above 36.5°C and below 37.5°C. But when Wunderlich's articles in German were translated into English, reaching a larger audience, the temperature was converted from centigrade to Fahrenheit and became 98.6°F – inviting physicians to assume that the temperature had been measured to one tenth of a degree Fahrenheit rather than one degree centigrade. The implied precision was almost twenty times greater – but all that had actually changed was a conversion between two temperature scales.

So how could so large a dataset be wrong? When Dr Mackowiak discovered one of Carl Wunderlich's old thermometers in a medical museum, he was able to inspect it. He found that it was miscalibrated by two degrees centigrade, almost four degrees Fahrenheit. This error was partly offset by Dr Wunderlich's habit of taking the temperature of the armpit rather than carefully inserting the thermometer into one of the bodily orifices conventionally used in modern times. You can take a million temperature readings, but if your thermometer is broken and you're poking around in armpits, then your results will be a precise estimate of the wrong answer. The old cliché of 'garbage in, garbage out' remains true no matter how many scraps of garbage you collect.

As we saw in the last chapter, the modern version of this old problem is an algorithm that has been trained on a systematically biased dataset. It's surprisingly easy for such problems to be overlooked. In 2014, Amazon, one of the most valuable companies in the world, started using a data-driven algorithm to sift résumés, hoping that the computer would find patterns and pick out the very best people, based on their similarity to previous successful applicants. Alas, previous successful applicants were disproportionately men. The algorithm then did what algorithms do: it spotted the pattern and ran with it. Observing that men had in the past been preferred, it concluded that men were preferable. The algorithm penalised the word 'women's' as in 'Under-21 Women's Soccer International' or 'Women's Chess Club Captain'; it downgraded certain all-women's colleges. Amazon abandoned the algorithm in 2018; it's unclear exactly how much influence it had in making decisions, but Amazon admitted that its recruiters had been looking at the algorithm's rankings.

Remember the 'Math is Racist' headline? I'm fairly

confident that maths isn't racist. Neither is it misogynistic, or homophobic, or biased in other ways. But I'm just as confident that some humans are. And computers trained on our own historical biases will repeat those biases at the very moment we're trying to leave them behind us.[14]

I hope I've persuaded you that we shouldn't be too eager to entrust our decisions to algorithms. But I don't want to overdo the critique, because we don't have some infallible alternative way of making decisions. The choice is between algorithms and humans. Some humans are prejudiced. Many humans are frequently tired, harassed and overworked. And all humans are, well, human.

In the 1950s, the psychologist Paul Meehl investigated whether the most basic of algorithms, in the form of uncomplicated statistical rules, could ever outperform expert human judgement. For example, a patient arrives at hospital complaining of chest pains. Does she have indigestion, or is she suffering from a heart attack? Meehl compared the verdicts of experienced doctors with the result of working through a brief checklist. Is chest pain the main symptom? Has the patient had heart attacks in the past? Has the patient used nitroglycerin to relieve chest pain in the past? What quantifiable patterns are shown on the cardiogram?[15] Disconcertingly, that simple decision-tree got the right diagnosis more often than the doctors. And it was not the only example. Remarkably often, Meehl found, experts fared poorly when compared with simple checklists. Meehl described his *Clinical vs. Statistical Prediction* as 'my disturbing little book'.[16]

So, to be fair, we should compare the fallibility of today's algorithms with that of the humans who would otherwise be making the decisions. A good place to start is with an example from Hannah Fry's book *Hello World*.

The story starts during the London riots of 2011. Initially a protest against police brutality, demonstrations turned into violent riots as order broke down each evening across the city, and in several other cities across the country. Shops would close in the early afternoon and law-abiding citizens would hurry home, knowing that opportunistic troublemakers would be out as the light faded. In three days of trouble, more than a thousand people were arrested.

Among their number were Nicholas Robinson and Richard Johnson. Robinson was walking through the chaos and helped himself to a pack of bottled water from a shattered London supermarket. Johnson drove to a gaming store, put on a balaclava, and ran in to grab an armful of computer games. Johnson's theft was of higher value and was premeditated rather than on the spur of the moment. And yet it was Robinson who received a six-month sentence, while Johnson's conviction earned him no jail time at all. No algorithm could be blamed for the difference; human judges were the ones handing out those sentences, and the disparity seems bizarre.

It's always possible that each judge made the right decision, based on some subtle detail of the case. But the most plausible answer for the inconsistent treatment of the two men was that Robinson was sentenced just two weeks after the riots, at a time when nerves were jangling and the fabric of civilisation seemed easily ripped. Johnson was sentenced months later, when the memory of the riots was fading and people were asking themselves what all the fuss had been about.[17]

Would a data-driven computer program have tuned out the mood music and delivered fairer sentences? It's impossible to know – but quite possibly, yes. There is ample evidence that human judges aren't terribly consistent. One way to test this is to show hypothetical cases to various judges and see if they reach different conclusions. They do. In one British study from 2001, judges were asked for judgements on a variety of cases; some of the cases (presented a suitable distance apart to disguise the subterfuge) were simply repeats of earlier cases, with names and other irrelevant details changed. The judges didn't even agree with their own previous judgement on the identical case. That is one error that we can be fairly sure a computer would not make.[18]

A more recent study was conducted in the United States, by economist Sendhil Mullainathan and four colleagues. They analysed over 750,000 cases in New York City between 2008 and 2013 – cases in which someone had been arrested and the decision had to be taken as to whether to release the defendant, or to detain him or her, or to set a cash bail that had to be posted to secure release. The researchers could then see who had gone on to commit further crimes. They then used a portion of these cases (220,000) to train an algorithm to decide whether to release, detain or set bail. And they used the remaining cases to check whether the algorithm had done a good job or not, relative to human judges.[19]

The humans did not do well. The researchers' algorithm could have reduced crime-while-on-release by almost 25 per cent by jailing a better-selected group of defendants. Alternatively, they could have jailed 40 per cent fewer people without any increase in crime. Thousands of crimes could have been prevented, or thousands of people released pending trial, purely as a result of the algorithm outperforming the human judges.

One important error that the judges make is what legal scholar Cass Sunstein calls 'current offence bias' – that is, when they make decisions about bail they focus too much on the specific offence the defendant has been accused of. Defendants whose track record suggests they're a high risk are treated as low risk if they're accused of a minor crime, and defendants whose track record suggests they're low risk are treated as high risk if the current offence is serious. There's valuable information here that the algorithm puts to good use, but the human judges – for all their intelligence, experience and training – tend to overlook.

This seems to be how we humans operate. Consider the way I described the cases of Nicholas Robinson and Richard Johnson: I told you about the offences in question, nothing at all about Robinson and Johnson. It just seemed reasonable to me – and perhaps to you – to tell you all about the short term, about the current offence. An algorithm would have used more information if more information had been available. A human might not.

Many people have strong intuitions about whether they would rather have a vital decision about them made by algorithms or humans. Some people are touchingly impressed by the capabilities of the algorithms; others have far too much faith in human judgement. The truth is that sometimes the algorithms will do better than the humans, and sometimes they won't. If we want to avoid the problems and unlock the promise of big data, we're going to need to assess the performance of the algorithms on a case-by-case basis. All too often, this is much harder than it should be.

Consider this scenario. The police, or social services, receive a call from someone – a neighbour, a grandparent, a doctor, a teacher – who is worried about the safety of a child. Sometimes the child will genuinely be in danger; sometimes the caller will be mistaken, or over-anxious, or even malicious. In an ideal world, we'd take no chances and send a blue-lighted vehicle around immediately to check what's going on. But we don't have enough resources to do this in every case – we have to prioritise. The stakes could hardly be higher: official figures in the United States show that 1670 children died in 2015 of abuse or neglect. That's a horrific number, but a tiny fraction of the 4 million times someone calls to report their concerns about a child.

Which reports need to be followed up, and which can be reasonably ignored? Many police and social services use algorithms to help make that decision. The state of Illinois introduced just such an algorithm, called Rapid Safety Feedback. It analysed data on each report, compared them to the outcomes of previous cases, and produced a percentage prediction of the child's risk of death or serious harm.

The results were not impressive. The *Chicago Tribune* reported that the algorithm gave 369 children a 100 per cent chance of serious injury or death. No matter how dire the home environment, that degree of certitude seems unduly pessimistic. It could also have grave implications: a false allegation of child neglect or abuse could have terrible consequences for the accused and the child alike.

But perhaps the algorithm erred on the side of caution, exaggerating the risk of harm because it was designed not to miss a single case? No: in some terrible cases toddlers died after being given a percentage risk too low to justify follow-up. In the end, Illinois decided the technology was useless, or worse, and stopped using it.[20]

The moral of this story isn't that algorithms shouldn't be used to assess reports about vulnerable children. Someone or something must make the decision on which cases to follow up. Mistakes are inevitable, and there's no reason – in principle – why some other algorithm might not make fewer mistakes than a human call handler.[21] The moral is that we know about the limitations of this particular algorithm only because it spat out explicit numbers that were obviously absurd.

'It's good they gave numerical probabilities, as this supplies the loud siren that makes us realize that these numbers are bad,' explains statistician Andrew Gelman. 'What would be worse is if [the algorithm] had just reported the predictions as "high risk", "mid risk" and "low risk".' The problems might then never have come to light.[22]

So the problem is not the algorithms, or the big datasets. The problem is a lack of scrutiny, transparency and debate. And the solution, I'd argue, goes back a very long time.

In the mid-seventeenth century, a distinction began to emerge between alchemy and what we'd regard as modern science. It is a distinction that we need to remember if we are to flourish in a world of big data algorithms.

In 1648, Blaise Pascal's brother-in-law, at the urging of the great French mathematician, conducted a celebrated experiment. In the garden of a monastery in the little city of Clermont-Ferrand, he took a tube filled with mercury, slid its open end into a bowl full of the liquid metal, and lifted it to a vertical position, jutting above the surface with the end submerged. Some of the mercury immediately drained into the bowl, but some did not. There was a column 711 millimetres high in the tube, and above it a space containing – what? Air? A vacuum? A mysterious ether?[23]

This was only the first stage of the experiment Pascal had proposed, and it wasn't unprecedented. Gasparo Berti had done something similar in Rome with water – although with water, the glass tube needs to be more than 10 metres long, and making one was no easy task. Evangelista Torricelli, a student of Galileo, was the man who had the idea of using mercury instead, which requires a much shorter tube.

Pascal's idea – or perhaps it was his friend René Descartes, since both of them claimed the credit – was to repeat the experiment at altitude. And so it was Pascal's brother-in-law who had the job of lugging fragile glass tubes and several kilograms of mercury up to the top of Puy de Dôme, a striking dormant volcano in the heart of France, more than a kilometre above Clermont-Ferrand. At the top of the mountain, the mercury rose not 711 millimetres but just 627. Halfway down the mountain, the mercury column was longer than at the summit, but shorter than down in the garden. The next day, the column was measured at the top of Clermont-Ferrand's cathedral. It was 4 millimetres shorter there than in the monastery garden. Pascal had invented what we now call the barometer – and simultaneously, the altimeter, a device that measured air pressure and, indirectly, altitude. In 1662, just fourteen years later, Robert Boyle formulated his famous gas law, describing the relationship between pressure and the volume of a gas. This was a rapid and rather modern advance in the state of scientific knowledge.

Yet it was taking place alongside the altogether more ancient practice of alchemy, the quest to find a way to turn base metals into gold and to produce an elixir of eternal life. These goals are, as far as we know, as near to impossible as makes no difference* – but if alchemy had been conducted using scientific methods one might still have expected all the alchemical research to produce a rich seam of informative failures, and a gradual evolution into modern chemistry.

That's not what happened. Alchemy did not evolve into chemistry. It stagnated, and in due course science elbowed it to one side. For a while the two disciplines existed in parallel. So what distinguished them?

---

* A particle accelerator will turn base metals into gold, although not cheaply. In 1980, researchers bombarded the faintly lead-like metal bismuth and created a few atoms of gold. The cost was a less-than-economical rate of one quadrillion dollars an ounce. We are yet to discover an elixir of eternal life.

Of course, modern science uses the experimental method, so clearly demonstrated by Pascal's hard-working brother-in-law, by Torricelli, Boyle and others. But so did alchemy. The alchemists were unrelenting experimenters. It's just that their experiments yielded no information that advanced the field as a whole. The use of experiments does not explain why chemistry flourished and alchemy died.

Perhaps, then, it was down to the characters involved? Perhaps the great early scientists such as Robert Boyle and Isaac Newton were sharper, wiser, more creative men than the alchemists they replaced? This is a spectacularly unpersuasive explanation. Two of the leading alchemists of the 1600s were Robert Boyle and Isaac Newton. They were energetic, even fervent, practitioners of alchemy, which thankfully did not prevent their enormous contributions to modern science.[24]

No – the alchemists were often the very same people using the same experimental methods to try to understand the world around them. What accounts for the difference, says David Wootton, a historian of science, is that alchemy was pursued in secret while science depended on open debate. In the late 1640s, a small network of experimenters across France, including Pascal, worked simultaneously on vacuum experiments. At least a hundred people are known to have performed these experiments between Torricelli's in 1643 and the formulation of Boyle's Law in 1662. 'These hundred people are the first dispersed community of experimental scientists,' says Wootton.[25]

At the centre of the web of knowledge was Marin Mersenne – a monk, a mathematician, and a catalyst for scientific collaboration and open competition. Mersenne was friends with Pascal and Descartes, along with thinkers from Galileo to Thomas Hobbes, and would make copies of the letters he received and circulate them to others whom he thought would be interested. So prolific was his correspondence that he became known as 'the post-box of Europe'.[26]

Mersenne died in 1648, less than three weeks before the experiment on Puy de Dôme, but his ideas about scientific collaboration lived on in the form of the Royal Society in London (established 1660) and the French Academy of Sciences (established 1666), both along distinctly Mersennian lines. One of the virtues of the new approach, well understood at the time, was reproducibility – which, as we saw in the fifth chapter, is a vital check on both fraud and error. The Puy de Dôme experiment could be and was repeated anywhere there was a hill or even a tall building. 'All the curious can test it themselves whenever they like,' wrote Pascal. And they did.

Yet while the debate over vacuums, gases and those tubes of mercury was being vigorously carried out through letters, publications and meetings at Mersenne's home in Paris, alchemical experiments were conducted in secret. It isn't hard to see why: there is no value to turning lead into gold if everyone knows how to do it. No alchemist wanted to share his potentially instructive failures with anyone else.

The secrecy was self-perpetuating. One of the reasons that alchemy lasted so long, and that even brilliant scholars such as Boyle and Newton took it seriously, was the assumption that alchemical problems had been solved by previous generations, but kept secret and then lost. When Newton famously declared 'if I have seen further it is by standing on the shoulders of giants', this was true only of his scientific work. As an alchemist, he stood on nobody's shoulders and saw little.

When Boyle did try to publish some of his findings, and seek out other alchemists, Newton warned him to stop and instead maintain 'high silence'. And as it became clear that the newly open scientific community was making rapid progress, alchemy itself became discredited within a generation. In short, says Wootton,

> What killed alchemy was the insistence that experiments must be openly reported in publications which presented a clear account of what had happened, and they must then be replicated, preferably before independent witnesses. The alchemists had pursued a secret learning ... some parts of that learning could be taken over by ... the new chemistry, but much of it had to be abandoned as incomprehensible and unreproducible. Esoteric knowledge was replaced by a new form of knowledge which depended both on publication and on public or semi-public performance.[27]

Alchemy is not the same as gathering big datasets and developing pattern-recognising algorithms. For one thing, alchemy is impossible, and deriving insights from big data is not. Yet the parallels should also be obvious. The likes of Google and Target are no more keen to share their datasets and algorithms than Newton was to share his alchemical experiments. Sometimes there are legal or ethical reasons – if you're trying to keep your pregnancy a secret, you don't want Target publicly disclosing your folic acid purchases – but most obviously the reasons are commercial. There's gold in the data that Amazon, Apple, Facebook, Google and Microsoft have about us. And that gold will be worth a lot less to them if the knowledge that produces it is shared with everyone.

But just as the most brilliant thinkers of the age failed to make progress while practising in secret, secret algorithms based on secret data are likely to lead to missed opportunities for improvement. Again, it hardly matters much if Target is missing out on a slightly more effective way to target onesie coupons. But when algorithms are firing capable teachers, directing social services to the wrong households, or downgrading job applicants who went to women's colleges, we need to be able to subject them to scrutiny.

But how?

One approach is that used by a team of investigative journalists at ProPublica, led by Julia Angwin. Angwin's team wanted to scrutinise a widely used algorithm called COMPAS (Correctional Offender Management Profiling for Alternative Sanctions). COMPAS used the answers to a 137-item questionnaire to assess the risk that a criminal might re-offend. But did it work? And was it fair?

It wasn't easy to find out. COMPAS is owned by a company, Equivant (formerly Northpointe), which is under no obligation to share the details of how it works. And so Angwin and her team had to judge it by analysing the results, laboriously pulled together from Broward County in Florida, a state that has strong transparency laws.

Here's an edited account of how the ProPublica team went about their work:

> Through a public records request, ProPublica obtained two years' worth of COMPAS scores from the Broward County Sheriff's Office in Florida. We received data for all 18,610 people who were scored in 2013 and 2014 ... Each pretrial defendant received at least three COMPAS scores: 'Risk of Recidivism,' 'Risk of Violence' and 'Risk of Failure to Appear.' COMPAS scores for each defendant ranged from 1 to 10, with ten being the highest risk. Scores 1 to 4 were labeled by COMPAS as 'Low'; 5 to 7 were labeled 'Medium'; and 8 to 10 were labeled 'High.' Starting with the database of COMPAS scores, we built a profile of

> each person's criminal history, both before and after they were scored. We collected public criminal records from the Broward County Clerk's Office website through April 1, 2016. On average, defendants in our dataset were not incarcerated for 622.87 days (sd: 329.19). We matched the criminal records to the COMPAS records using a person's first and last names and date of birth ... We downloaded around 80,000 criminal records from the Broward County Clerk's Office website.[28]

And so it continues. This was painstaking work.

Eventually, ProPublica published their conclusions. Although the COMPAS algorithm did not use an offender's race as a predictor, it nevertheless was producing racially disparate results. It tended to produce false positives for black offenders (predicting that they would re-offend, but then they did not) and false negatives for white offenders (predicting that they would not re-offend, but then they did).

That sounds very worrying: racial discrimination is both immoral and illegal when coming from a human; we shouldn't tolerate it if it emerges from an algorithm.

But then four academic researchers, Sam Corbett-Davies, Emma Pierson, Avi Feller and Sharad Goel, pointed out that the situation wasn't so clear-cut.[29] They used the data laboriously assembled by ProPublica to show that the algorithm was fair by another important metric, which was that if the algorithm gave two criminals – one black, one white – the same risk rating, then the actual risk that they re-offended was the same. In that important respect the algorithm was colour-blind.

What's more, the researchers showed that it was impossible for the algorithm to be fair in both ways simultaneously. It was possible to craft an algorithm that would give an equal rate of false positives for all races, and it was possible to craft an algorithm where the risk ratings matched the re-offending risk for all races, but it wasn't possible to do both at the same time: the numbers just couldn't be made to add up.

The only way in which an algorithm could be constructed to produce equal results for different groups – whether those groups were defined by age, gender, race, hair colour, height or any other criterion – would be if the groups otherwise behaved and were treated identically. If they moved through the world in different ways, the algorithm would, inevitably, violate at least one criterion of fairness when evaluating them. That is true whether or not the algorithm was actually told their age, gender, race, hair colour or height. It would also be true of a human judge; it's a matter of arithmetic.

Julia Dressel and Hany Farid, also computer scientists, observed this debate over whether COMPAS was producing results with a racial bias. They thought there was something missing. 'There was this underlying assumption in the conversation that the algorithm's predictions were inherently better than human ones,' Dressel told the science writer Ed Yong, 'but I couldn't find any research proving that.'[30]

Thanks to ProPublica's spade-work, Dressel and Farid could investigate the question for themselves. Even if COMPAS itself was a secret, ProPublica had published enough of the results to allow it to be meaningfully tested against other benchmarks. One was a simple mathematical model with just two variables: the age of the offender and the number of previous offences. Dressel and Farid showed that the two-variable model was just as accurate as the much-vaunted 137-variable COMPAS model. Dressel and Farid also tested COMPAS predictions against the judgement of ordinary, non-expert humans who were shown just seven pieces of information about each offender and asked to predict whether he or she would re-offend within two years. The average of a few of these non-expert predictions outperformed the COMPAS algorithm.

This is striking stuff. As Farid commented, a judge might be impressed if told that a data-driven algorithm had rated a person as high-risk, but would be far less impressed if told, 'Hey, I asked twenty random people online if this person will recidivate and they said yes.'[31]

Is it too much to ask COMPAS to beat the judgement of twenty random people from the internet? It doesn't seem to be a high bar; nevertheless COMPAS could not clear it.[32]

Demonstrating the limitations of the COMPAS algorithm wasn't hard once the ProPublica data on COMPAS's decision-making had been released to allow researchers to analyse and debate them. Keeping the algorithms and the datasets under wraps is the mindset of the alchemist. Sharing them openly so they can be analysed, debated and – hopefully – improved on? That's the mindset of the scientist.

Listen to the speeches of traditional centre-ground politicians, or read media commentary, and it is common to encounter a view such as 'levels of trust are declining', or that 'we need to rebuild trust'. Baroness Onora O'Neill, who has become an authority on the topic, argues that such hand-wringing reflects sloppy thinking. She argues that we don't and shouldn't trust in general: we trust specific people or institutions to do specific things. (For example: I have a friend I'd never trust to post a letter for me – but I'd gladly trust him to take care of my children.) Trust should be discriminating: ideally we should trust the trustworthy, and distrust the incompetent or malign.[33]

Just like people, algorithms are neither trustworthy nor untrustworthy as a general class. Just as with people, rather than asking 'Should we trust algorithms?' we should ask 'Which algorithms can we trust, and what can we trust them to do?'

Onora O'Neill argues that if we want to demonstrate trustworthiness, we need the basis of our decisions to be 'intelligently open'. She proposes a checklist of four properties that intelligently open decisions should have. Information should be *accessible*: that implies it's not hiding deep in some secret data vault. Decisions should be *understandable* – capable of being explained clearly and in plain language. Information should be *usable* – which may mean something as simple as making data available in a standard digital format. And decisions should be *assessable* – meaning that anyone with the time and expertise has the detail required to rigorously test any claims or decisions if they wish to.

O'Neill's principles seem like a sensible way to approach algorithms entrusted with life-changing responsibilities, such as whether to release a prisoner, or respond to a report of child abuse. It should be possible for independent experts to get under the hood and see how the computers are making their decisions. When we have legal protections – for example, forbidding discrimination on the grounds of race, sexuality or gender – we need to ensure that the algorithms live up to the same standards we expect from humans. At the very least that means the algorithm needs to be scrutable in court.

Cathy O'Neil, author of *Weapons of Math Destruction*, argues that data scientists should – like doctors – form a professional society with a professional code of ethics. If nothing else, that would provide an outlet for whistle-blowers, 'so that we'd have someone to complain to when our employer (Facebook, say) is asking us to do something we suspect is unethical or at least isn't up to the standards of accountability that we've all agreed to'.[34]

Another parallel with the practice of medicine is that important algorithms should be tested using randomised controlled trials. If an algorithm's creators claim that it will sack the right teachers, or recommend bail for the right criminal suspects, our response should be, 'prove it'. The history of medicine teaches us that plausible-sounding ideas can be found wanting when subjected to a fair test. Algorithms aren't medicines, so simply cloning an organisation such as the US Food and Drug Administration wouldn't work; we'd need to run the trials over faster timelines and take a different view of what informed consent looked like. (Clinical trials have high standards for ensuring that people consent to participate; it's not so clear how those standards would apply to an algorithm that rates teachers – or criminal suspects.) Still, anyone who is confident of the effectiveness of their algorithm should be happy to demonstrate that effectiveness in a fair and rigorous test. And vital institutions such as schools and courts shouldn't be willing to use those algorithms on a large scale unless they've proved themselves.

Clearly, not all algorithms raise such weighty concerns. It wouldn't obviously serve the public interest to force Target to let researchers see how they decide who receives onesie coupons. We need to look on a case-by-case basis. What sort of accountability or transparency we want depends on what problem we are trying to solve.

We might, for example, want to distinguish YouTube's algorithm for recommending videos from Netflix's algorithm for recommending movies. There is plenty of disturbing content on YouTube, and its recommendation engine has become notorious for its apparent tendency to suggest ever more fringey and conspiratorial videos. It's not clear that the evidence supports the idea that YouTube is an engine of radicalisation, but without more transparency it's hard to be sure.[35]

Netflix illustrates a different issue: competition. Its recommendation algorithm draws on a huge, secret dataset of which customers have watched which movies. Amazon has a similar, equally secret dataset. Suppose I'm a young entrepreneur with a brilliant idea for a new kind of algorithm to predict which movies people will like based on their previous viewing habits. Without the data to test it on, my brilliant idea can never be realised. There's no particular reason for us to worry about how the Amazon and Netflix algorithms work, but is there a case for forcing them to make public their movie-viewing datasets, to unleash competition in algorithm design that might ultimately benefit consumers?

One concern is immediately apparent – privacy. You might think that was an easy problem to solve: just remove the names from records and the data are anonymous! Not so fast: with a rich dataset, and by cross-referencing with other datasets, it is often surprisingly easy to figure out who Individual #961860384 actually is. Netflix once released an anonymised dataset to researchers as part of a competition to find a better recommendation algorithm. Unfortunately, it turned out that one of their customers had posted the same review of a family movie on Netflix and, under her real name, on the Internet Movie Database website. Her no-longer-anonymous Netflix reviews revealed that she was attracted to other women – something she preferred to keep secret.[36] She sued the company for 'outing' her; it settled on undisclosed terms.

Still, there are ways forward. One is to allow secure access to certified researchers. Another is to release 'fuzzy' data where all the individual details are a little bit off, but rigorous conclusions can still be drawn about populations as a whole. Companies such as Google and Facebook gain an enormous competitive advantage from their datasets: they can nip small competitors in the bud, or use data from one service (such as

Google Search) to promote another (such as Google Maps or Android). If some of that data were made publicly available, other companies would be able to learn from it, produce better services, and challenge the big players. Scientists and social scientists could learn a lot, too; one possible model is to require private 'big data' sets to be published after a delay and with suitable protections of anonymity. Three-year-old data are stale for many commercial purposes but may still be of tremendous scientific value.

There is a precedent for this: patent holders must publish their ideas in order to receive any intellectual property protection; perhaps a similar bargain could be offered to, or imposed on, private holders of large datasets.

'Big data' is revolutionising the world around us, and it is easy to feel alienated by tales of computers handing down decisions made in ways we don't understand. I think we're right to be concerned. Modern data analytics can produce some miraculous results, but big data is often less trustworthy than small data. Small data can typically be scrutinised; big data tends to be locked away in the vaults of Silicon Valley. The simple statistical tools used to analyse small datasets are usually easy to check; pattern-recognising algorithms can all too easily be mysterious and commercially sensitive black boxes.

I've argued that we need to be sceptical of both hype and hysteria. We should ask tough questions on a case-by-case basis whenever we have reason for concern. Are the underlying data accessible? Has the performance of the algorithm been assessed rigorously – for example, by running a randomised trial to see if people make better decisions with or without algorithmic advice? Have independent experts been given a chance to evaluate the algorithm? What have they concluded? We should not simply trust that algorithms are doing a better job than humans, and nor should we assume that if the algorithms are flawed, the humans would be flawless.

But there is one source of statistics that, at least for the citizens of most rich countries, I think we should trust more than we do. And it is to this source that we now turn.

RULE EIGHT

Don't take statistical bedrock for granted

'What do you base your facts on?'
  'Statistics from the International Monetary Fund and the United Nations, nothing controversial. These facts are not up for discussion. I am right, and you are wrong.'

—HANS ROSLING[1]

Monday, 9 October 1974. The place: Washington DC, near the picturesque tidal basin – a quiet, leafy sanctuary not far from the White House. The time: two o'clock in the morning. A car is weaving around in the darkness, at speed, with its headlights off. The police pull the car over, at which point a flamboyantly dressed woman with two black eyes jumps out of the passenger's side, runs down the road yelling alternately in English and Spanish, and leaps into the water. The police pull her out, and she tries to jump in again, at which point they handcuff her. At the wheel is an elderly fellow with broken glasses and minor cuts to his face. He is steaming drunk.[2]

Just another night in DC, perhaps. Except the woman, Annabelle Battistella, was better known as Fanne Foxe, The Argentine Firecracker, an erotic dancer at the Silver Slipper nightclub. And the man was one of the most powerful men in the United States: Wilbur Mills, an Arkansas congressman since the 1930s, who as the long-serving chair of the House Ways and Means Committee effectively had veto power over most legislation. These were, however, deferential times. The police offered to drive Mr Mills home to his wife in his own car, and he was re-elected by voters just a few weeks later.

But hard on the heels of that electoral triumph, Mr Mills – drunk again – appeared on stage with Foxe in the middle of her act, received a peck on the cheek, and made his exit, stage left. To be caught cavorting with a stripper once might be considered a misfortune. To do it twice suggested carelessness. His colleagues had a quiet word. Wilbur Mills stepped down from the Ways and Means Committee and joined Alcoholics Anonymous. Fanne Foxe rebranded herself 'The Tidal Basin Bombshell', wrote a tell-all memoir, and eventually retired into obscurity.[3]

To most people, this story might be dimly remembered as America's third most spectacular sex scandal. But in my home country of Nerdland, it has another significance. At the time, Congress was deadlocked over a putative new agency, the Congressional Budget Office, which would provide advice to Congress about the budgetary costs of different policy proposals. One congressional dinosaur in particular objected to plans to appoint a woman as its director. But Wilbur Mills's resignation triggered the usual game of musical chairs, the indirect consequence of which was that the deadlock was broken. The Congressional Budget Office was duly established, and with the dinosaur ambling off to graze elsewhere on Capitol Hill, there was no obstacle to its first director being the woman that every sensible person would have wanted: Alice Rivlin. Forty years later, she reflected, 'I owed my job to Fanne Foxe'.[4]

After this strange beginning, Alice Rivlin went on to lead the Congressional Budget Office to glory.* The CBO had been established by Congress to serve as a counterweight to what was seen as Richard Nixon's overreaching and over-mighty presidency. Congress saw the value of having better statistics, and more analysis of policy issues. But Rivlin interpreted this role in a particular way: rather than churning out talking points for the majority party or running statistical errands for the powerful chairs of congressional committees, she would supply impartial, high-quality information and analysis to Congress as a whole. In the judgement of one academic, the CBO duly became 'one of the most influential and well-regarded institutions in Washington ... the authoritative source of information on the budget and the economy'.[5]

Alice Rivlin's deputy and later one of her successors, Robert Reischauer, described the CBO as

> basically a manhole in which Congress would have a bill or something, and it would lift up the manhole cover and put the bill down it, and you would hear grinding noises, and twenty minutes later a piece of paper would be handed up, with the cost estimate, the answer, on it. No visibility, [just] some kind of mechanism below the ground level doing this ... noncontroversial, the way the sewer system is.[6]

---

* She also became president of the American Economic Association, deputy at the Federal Reserve, and head of the Office of Management and the Budget. She also 'saved Washington', as the *Washington Post* put it, by helping guide the city of Washington DC out of its own local fiscal crisis in the 1990s. Given this range of high-level positions, one colleague called her a 'decathlete' of public service.

The analogy is apt, and not just because sewers are invisible and uncontroversial. Independent statistical agencies, like sewers, are an essential part of modern life. Like sewers, we tend to take them for granted until something goes wrong. And like sewers, they can suffer badly from neglect – or because someone tries to force something unsuitable through them for their own selfish or foolish reasons.

The official statistics and analyses produced by organisations such as the CBO are more important than we might think, and more useful in the everyday lives of ordinary citizens. They are also under threat – and we should defend them. They should not have to depend on twists of fate involving drunk congressmen and strippers.

The CBO was established, remember, with Richard Nixon in mind. But Nixon had resigned before the CBO began operations, and the first President to object to what the CBO was doing was not a Republican like Nixon, but a Democrat: Jimmy Carter. With oil prices spiking in the late 1970s, President Carter had ambitious goals to improve America's energy efficiency. Alice Rivlin's CBO team evaluated the proposals and judged that they wouldn't work as well as Carter hoped.

'It made the Carter Administration unhappy,' Dr Rivlin later recalled. The House Speaker, also a Democrat, wasn't happy either. 'He was fighting for the legislation and the CBO wasn't helping.'[7]

No. It wasn't helping. This was exactly the point: Alice Rivlin knew that the value of the CBO would lie in being impartial rather than in serving up propaganda for the party in power. It wasn't long before the party in power was the Republicans again, and it was their turn for their grand claims to smack into the unyielding reality of the CBO's independent opinion. In 1981, the CBO argued that the budget deficit was likely to be far higher than the Reagan White House projected. President Reagan called the CBO numbers 'phony'.

In 1983, Alice Rivlin left the CBO after eight years in charge. Successive administrations continued to put pressure on it – in the 1990s, for example, leading Democrats wanted the CBO to produce a more flattering analysis of President Clinton's health care reforms – and it continued to assert its independence.[8] The CBO certainly isn't perfect: much of its task is to make projections of the future gap between spending and tax revenue, and – as we'll discuss in the tenth chapter – such economic forecasts are hard to make; official agencies often get them wrong. The important point, though, is that they don't make politically expedient errors, systematically warping their forecasts to fit a political agenda. Evaluations of the CBO[9] have tended to find that it produces forecasts that are as accurate as we might reasonably hope, and – crucially – unbiased.*

In the UK, the Office for Budget Responsibility (OBR) performs a similar role to the CBO. It was established as an independent agency only in 2010. Forecasts of spending, tax receipts and other economic variables had previously been made by the Treasury, where officials are more directly answerable to politicians. That enables us to make an interesting comparison: are the OBR forecasts better? It turns out they are, substantially so.[10] That's encouraging for the OBR's reputation and future work, but it also suggests there

* For example, a peer-reviewed study by two academics, published in 2000, found that Republican administrations tended to produce forecasts that were too worried about high inflation, while Democratic administration forecasts were too pessimistic about unemployment. CBO forecasts showed neither bias and were more accurate overall.

was previously a problem – that before 2010, the Treasury's economists had been routinely shaping their forecasts to please their political overlords.

The CBO and OBR are far from the only kinds of statistical agencies that need to assert their political independence. While they project the future impacts of proposed tax or spending changes, many other agencies describe current realities. There are censuses, which try to estimate how many people live in different parts of the country, along with some basic information about those people. There are economic statistics – measuring inflation, unemployment, economic growth, trade and inequality. There are social statistics – measuring crime, education, access to housing, migration and well-being. There are studies of particular industries, or of issues such as environmental pollution.

Every country has its own arrangements for putting together these official statistics. In the UK, many are produced and published by one organisation, the Office for National Statistics. In the United States, the task is spread across a range of agencies including the Bureau of Economic Analysis, the Bureau of Labor Statistics, the Census Bureau, the Federal Reserve, the Department of Agriculture and the Energy Information Administration.

How useful is all this counting and measurement? Very useful indeed; it's hard to overstate how useful. The numbers produced by such agencies are a nation's statistical bedrock. When journalists, think-tanks, academics and fact-checkers want to know what is going on, their analysis usually rests, either directly or ultimately, on this bedrock. I'll have more to say about the costs and varied benefits of producing professional and impartial official statistics later in the chapter. But perhaps the most vivid argument for their value is to look at attempts to distort, discredit or suppress them.

As a candidate for US President in 2016, Donald Trump faced a problem. His campaign wanted to claim that the American economy was broken, but official statistics showed that the unemployment rate was very low – below 5 per cent and falling. There could have been a thoughtful response to that – for example, that the unemployment rate doesn't measure the quality, security or earning power of jobs. But Mr Trump took the simpler path of repeatedly dismissing unemployment figures as 'phony' and 'total fiction' and claiming that the true rate was 35 per cent.

Simply inventing your own numbers is a tactic more often used by totalitarian dictators than by candidates for democratic election, but Mr Trump evidently figured it was a tactic that would be effective. And perhaps he was right. His supporters believed him: just 13 per cent of them trusted the economic data produced by the federal government, versus 86 per cent of those who voted for Hillary Clinton.[11]

As President, Mr Trump changed his mind. According to the official data, unemployment crept even lower after he had assumed office. Now, however, Mr Trump wished to get credit for this rather than to dismiss it. His spokesman Sean Spicer declared, with a straight face, 'I talked to the President prior to this, and he said to quote him very clearly. They may have been phony in the past, but it's very real now.' Amusing as this kind of shamelessness might be, it also carries a real risk – that Mr Trump's opponents will start to distrust official statistics just as much as his supporters do.[12]

If you grow tired of undermining trust in your own statistical agency when it isn't producing politically convenient figures, you could always attack the statistical agency of someone else. For example, after Germany's leader Angela Merkel took the politically risky step of welcoming almost a million refugees into the country in 2015, Donald Trump wanted to use Germany as a cautionary tale. 'Crime in Germany is way up' he tweeted in June 2018. Look at all the crimes those refugees were causing!

Unfortunately for President Trump, one group of people stood in the way to spoil his story: German statisticians. Their latest figures, a month before Trump's tweet, showed that not only was crime in Germany *not* 'way up', it was at its lowest level since 1992.[13] Unabashed, Trump had an answer. A few hours later, he tweeted that 'Crime in Germany is up 10% plus (officials do not want to report these crimes)'.[14]

The allegation is implausible. In part that's because the ministry in Germany responsible for putting together the police crime statistics was run by Horst Seehofer, an immigration hawk who in the same year threatened to resign if Germany's immigration policy wasn't tightened up: Mr Seehofer would hardly have wanted to pressure officials to hide uncomfortable truths about migration. It's also implausible because Germany has not acquired a reputation for political interference in statistics.

Sadly, that's not true for every country. Around the world, pressure to fiddle the figures is real and widespread – and the consequences for statisticians can be far more serious than grumbling from senior politicians.

In 2010, the economist Andreas Georgiou left a two-decade career at the International Monetary Fund, bringing his baby daughter with him from Washington DC to his home country, Greece. His mission was to run ELSTAT, Greece's new statistical agency.

At the time, Greece's statistics were in bad shape. They had never been well funded or well respected. When, in 2002, the economist Paola Subacchi visited the Greek statistical office she found it tucked away in a residential suburb of Athens, 'in a square of ordinary shops, and I had to hunt for a doorway in a 1950s apartment block that took me up some stairs to a dusty room with a handful of people. I can't remember seeing any computers. It was extraordinary, not a professional operation at all.'[15]

But when Georgiou arrived, there was more to worry about than dust and outdated technology. The entire world had reached the conclusion that you should trust Greek official statistics about as much as you should trust their giant wooden horses. Eurostat, the statistical office of the European Union, had repeatedly complained about the credibility and quality of the official Greek economic data. The European Commission issued a blistering report about them.[16]

The basic problem was that Greece was supposed to keep its government budget deficit at a modest level. The budget deficit is the amount the government borrows each year to cover any gap between what it spends and what it receives in taxes. One of the obligations that comes with membership of the Eurozone is for a country to keep its deficit below 3 per cent of gross domestic product, with some exemptions for various exceptional circumstances. (Economically speaking, it's not a very sensible rule – but that's another story for another book.) That target was onerous, so why not tweak the figures until all seemed well? One year the Greek accounts left out several billion euros of borrowing to pay for hospitals. Another year, they omitted a big chunk of the cost of the military. They also did a deal with the investment bank Goldman Sachs that effectively made borrowing look like a different kind of transaction, and thus not counting towards the deficit.[17]

In 2009, the shock of the global financial crisis was followed by the realisation that Greece had been underplaying its borrowing for years. Nobody believed its debts could be repaid. The EU and IMF stepped in with the customary mix of a bailout and some brutal austerity, and the Greek economy collapsed. Into this situation stepped Andreas Georgiou. He might not be able to rescue Greece's prosperity, but there was some hope that he would save the reputation of Greek official statistics.

Mr Georgiou's first priority was to look at the deficit figures for 2009, the most recent available. The initial forecast, from the Greek Ministry of Finance, had been 3.7 per cent of gross domestic product – not too far outside the EU's target, but unfortunately quite implausible. Even before Mr Georgiou's arrival the Greek authorities had revised that to a shocking 13.6 per cent. Eurostat were still unconvinced. Within a few months, Mr Georgiou published his conclusion: the deficit had actually been 15.4 per cent, a grimly large number. But it was, at least, believable – and Eurostat believed it.

It was then that Mr Georgiou's troubles began. First, there was an almighty row within ELSTAT. The police eventually realised that Mr Georgiou's email account had been hacked by his own deputy, ELSTAT's vice-president. Then the Greek Prosecutor of Economic Crimes began legal action against Mr Georgiou, accusing him of deliberately exaggerating Greece's deficit and causing immense damage to the Greek economy. Various other charges were added, including failing to allow ELSTAT's board to vote on what the deficit should be. (The idea that the size of Greece's budget deficit should be put to a vote seems more Eurovision than Eurostat.) The potential sentence for Mr Georgiou's 'crimes' was life imprisonment. The judicial system threw out the charges six times, but they were repeatedly reinstated by the Greek supreme court. Indeed, his convictions, acquittals and re-convictions have been so frequent that it is hard to have any confidence that any verdict will stick.[18] This is harassment worthy of a Kafka novel.

Of course it is possible that Georgiou really is a traitor. But it does not seem likely. Eighty former chief statisticians from around the world signed a letter protesting against his treatment, Eurostat repeatedly signed off on the quality of his work, and in 2018 he received a special commendation from a group of respected professional bodies including the International Statistical Association, the American Statistical Association and the Royal Statistical Society 'for his competency and strength in the face of adversity, his commitment to the production of quality and trustworthiness of official statistics and his advocacy for the improvement, integrity and independence of official statistics'.[19]

Andreas Georgiou is not the only statistician who has shown courage in adversity, as Graciela Bevacqua, a long-serving Argentine statistician, could attest. Argentina has long suffered from high inflation. The Argentine government, under a husband-and-wife pair of populist presidents Néstor Kirchner (president 2003-7) and Cristina Fernández de Kirchner (president 2007-15), decided to solve the problem not by reducing inflation but by changing the inflation statistics. Ms Bevacqua found herself receiving some discomfiting demands.

For example, she was instructed to round down all decimals in the monthly inflation figures – as though Argentine computers had run out of decimal points. That makes more difference than you might think, because each distortion compounds the earlier ones: compounding inflation of 1 per cent a month gives 12.7 per cent a year, while 1.9 per cent a month is 25.3 per cent a year. Funnily enough, official estimates of annual inflation in Argentina have tended to be close to the first number, and independent unofficial estimates have been closer to the second.

When Graciela Bevacqua produced a monthly figure of 2.1 per cent inflation at the beginning of 2007, her supervisors weren't happy. Hadn't they told her to produce a number below 1.5 per cent? They told her to take a vacation, then sacked her when she returned, transferring her from the statistical agency to a library and slashing her pay by two thirds. She resigned soon afterwards.[20]

With Ms Bevacqua out of the way – and having been made an example of – Argentina's official inflation numbers showed inflation of below 10 per cent. That's high by the standards of a developed country but still implausibly low. Most independent experts reckoned it was close to 25 per cent, and a group of those experts produced their own unofficial price index, advised by none other than Graciela Bevacqua – who was promptly fined $250,000 for false advertising.

As with Mr Georgiou, international observers stand behind Ms Bevacqua and her methods, and with a new government in Argentina it looks like she'll be OK. As for Mr Georgiou, he stuck it out for five years at ELSTAT then returned to the United States, leaving behind him an organisation with a credibility it never had before he arrived. He is most unlikely to go to prison, but other Greek statisticians will have noticed the way he was persecuted for trying to tell the truth about the statistics that were his responsibility. 'It will not be lost on them that their well-being – not only professional but personal – is at risk if they do the right thing and follow the law,' he told *Significance* magazine. He added that the Greek government was only damaging itself in the long run, by 'undermining the statistics which they themselves use. They're undermining the credibility of the country itself.' Meanwhile, the people who repeatedly understated Greece's deficit before the crisis seem to have escaped censure.[21]

Heroic as Andreas Georgiou and Graciela Bevacqua have shown themselves to be, we would be naive to assume that every statistician has their determination, or that every attempt to exert pressure comes to public attention. One respected statistician, Professor Denise Lievesley, told me that a fellow statistician from Africa had been told that if he didn't produce the numbers that his nation's president required, his children would be murdered. For understandable reasons, she didn't wish to identify him.[22] It would be equally understandable if he had decided to comply.

There are subtler ways to undermine the independence of official statisticians. In Tanzania, in late 2018, the government passed a law making criticism of official statistics a criminal offence, punishable with fines or a minimum of three years in prison. Candidates for the presidency there will think twice before following Mr Trump's example of calling the jobless figures 'phony'. But imprisoning anyone who finds fault with government statistics is not only an outrage against free speech, it will ensure that faults go uncorrected. Tanzania's move – which has been criticised by the World Bank – would be the perfect prelude to distorting its own statistics for political reasons.[23]

In India, Prime Minister Narendra Modi's government quietly stopped publishing data on unemployment in 2019. Mr Modi had made big promises about creating jobs, but in the run-up to that year's election (which he won comfortably) it began to look as though reality was going to prove embarrassing. The answer was simply to find an excuse to stop publishing, pending the arrival of 'improvements' in the data. One Indian expert explained to the *Financial Times* exactly what was going on: 'It's very clear that for a long time, the objective of the government has been to keep the picture fuzzy.'[24]

Even in countries with the most solid of reputations in Nerdland, serious conflicts can arise between the politicians and the statisticians. The Canadian statistical agency, Statistics Canada, has long been admired by statistical agencies around the world for its competence and independence – but the same qualities are not always appreciated closer to home. First the government under Prime Minister Stephen Harper (2006–15) tried to abolish the traditional census, replacing it with a voluntary survey – something that would have been cheaper and more convenient but massively less robust. The Chief Statistician, Munir Sheikh, made his objections very public and resigned.[25] The Harper government also wanted to move IT infrastructure to an organisation called Shared Services Canada; when the administration of the next Prime Minister, Justin Trudeau, pressed ahead with that plan, the *next* Chief Statistician, Wayne Smith, also resigned. He argued that if his data and computing power were being moved into another organisation, he could not guarantee the confidentiality of the statistics he was collecting. Nor could he be sure that Canadian statisticians would remain independent, since they could be squeezed or pressured by any government official with power over Shared Services Canada.

It's fair to say that Statistics Canada's reputation for robust independence has only been enhanced by these episodes. But there is a risk that if one side of the political spectrum is seen as hostile to the statisticians while the other side leaps to their defence, statistics itself becomes a partisan political issue. With that in mind, perhaps we should be reassured that when the last two Chief Statisticians of Canada resigned in protest, they did so under two different governments.[26]

*

In Puerto Rico, the government's response to troublesome statisticians was more radical: they attempted to disband entirely the statistical agency, PRIS, soon after the disastrous hurricane of September 2017. The ostensible reason was that PRIS was too expensive: its million-dollar budget could be better spent elsewhere.

That may not have been the real reason. You may recall that shortly after that hurricane, President Trump expressed gratitude that the death toll had been so small – sixteen or seventeen people, not a 'real tragedy' like the hurricane that had flooded New Orleans twelve years earlier. That was glib, but in line with the official death toll at the time – which later rose, but only to just over fifty. It seemed suspiciously low. Numerous independent researchers attempted to figure out their own estimates, to include not just the people who had been killed outright by the storm but those who had later died because of overstretched medical services, or because they were cut off from assistance by blocked roads and downed power lines. Alexis Santos was one of these researchers. He is a demographer at Penn State, and his Puerto Rican mother was on the island when the hurricane struck. Professor Santos put out an estimate that around a thousand people had died, directly or indirectly, as a result of the hurricane. It was big news in Puerto Rico. Even graver estimates were published later.

All of these estimates were built on demographic data from PRIS. PRIS itself, meanwhile, was suing the Puerto Rican health ministry in an effort to get accurate, timely information about the dead.[27] Given the embarrassment it was causing to the administration, perhaps the threat to disband PRIS was not entirely surprising.

Still, let's take the given reason at face value: is PRIS really worth its million-dollar budget? The question of how much value official statistics create is a valid one, and there are fewer attempts to quantify this than one might hope.

One cost-benefit exercise was conducted in the UK in the run-up to the 2011 census; it produced a long list of benefits from the census, everything from informing the debate over pension policy to ensuring that schools and hospitals were located in the right areas to enabling all sorts of other statistics to be calculated. After all, you can't produce any 'per capita' statistics – from crime to teen pregnancy to income to the unemployment rate – unless you know the population.

The analysts observed that 'statistics in themselves don't deliver benefits. It's the use of statistics that delivers benefits through better, quicker decisions by governments, companies, charities and individuals.'[28] That sounds plausible, and there are some surprising examples. London's Metropolitan Police, for example, used the census to identify streets with large numbers of elderly residents, and focused efforts on preventing fraudsters and burglars preying on vulnerable people. Everything from public health campaigns to nuclear disaster contingency plans depend on figuring out where everyone lives.

Disappointingly, the cost-benefit analysts shrugged their shoulders and declared themselves unable to put a value on all this, except to declare that it was obviously jolly useful. Still, they did find some benefits they judged to be quantifiable, and they pegged those as being worth £500 million a year – a bit less than £10 per UK resident. Since the census itself cost less than £500 million and lasts ten years, that suggests a tenfold return is a rather conservative estimate of the benefits.

Another attempt to tot up the value of official statistics was made in New Zealand, where the census, which cost NZ$200 million to conduct (about £100 million), was reckoned to have produced a benefit of at least a billion New Zealand dollars – a five-fold return. The study reckoned that refreshing the basic knowledge provided by the census – who lives where – produced a more accurate allocation of public spending on facilities such as hospitals and roads, and better-informed policy more generally.[29] Back in Puerto Rico, researchers pointed out ways in which PRIS had paid for itself, such as enabling the introduction of new systems to prevent fraud in collecting Medicare payments.[30]

But perhaps the strongest evidence that statistics are worthwhile is how cheap they are to collect, relative to the value of the decisions they inform. Consider the CBO: it advises Congress on $4 trillion worth of annual spending, on a budget of just $50 million a year. To put it another way, for every $80,000 the US government spends, one dollar funds the CBO to shed light on the other $79,999.[31] To justify its existence, the CBO would need to improve the effectiveness of government spending decisions by a mere 0.00125 per cent. It's hard to imagine how the CBO could fail to clear that bar.

Likewise, the million-dollar budget of PRIS sounds a lot more modest when you put it in the context of the Puerto Rican government's overall spending, which at nearly $10 billion is about 10,000 times larger. The UK's Office for National Statistics costs about £250 million a year – less than one pound for every £3000 the UK government spends. Between them, the thirteen principal statistical agencies in the US cost one dollar for every $2000 the US federal government spends.[32] If serious, independently gathered data improve government decision-making even by a tiny fraction, then these agencies are well worth the small sliver of public spending that is devoted to them.

*

Without statistics, then, governments would fumble in ignorance. But there is an intriguing counterargument, which is that governments are so reliably incompetent that giving them more information is risky: it will only encourage them.

One prominent advocate of this view was Sir John Cowperthwaite. Sir John was the financial secretary of Hong Kong throughout the 1960s, at a time when it was still under the control of the British – and when it was experiencing scorchingly rapid economic growth. Exactly how rapid was hard to say, because Sir John refused to collect basic information about Hong Kong's economy. The economist Milton Friedman, later to win the Nobel Memorial Prize in Economics, met Sir John at the time and asked him why. 'Cowperthwaite explained that he had resisted requests from civil servants to provide such data because he was convinced that once the data was published there would be pressure to use them for government intervention in the economy.'[33]

There was a logic to this. Hong Kong's rapid growth was partly thanks to an influx of immigrants from famine-struck communist China, but Cowperthwaite and Friedman also believed – with some reason – that it was flourishing thanks to a laissez-faire approach to policy. Cowperthwaite's government levied low taxes and provided very little in the way of public services. The private sector, he argued, would tend to solve people's problems more quickly and efficiently than the state. Why, then, collect data that would only encourage meddling from the authorities back in London? Cowperthwaite figured that the less London's politicians did, the better – and the less they knew, the less they would try to do.

Similarly, in his magisterial book *Seeing Like a State*, James C. Scott argues that the statistical information that states gather is flawed, missing the local details that matter. Imagine, say, a rural community in southeast Asia with complex customs regarding a piece of local land. Every household has some rights to farm the land, in rough proportion to its number of able-bodied members; then after each harvest, it becomes common land for grazing. Everyone can gather firewood, too, but the village baker and blacksmith are allowed to gather more. A surveyor from the new national land registry turns up, asking, 'Who owns this land?' Well – it's not so simple.

Now, it's one thing to be wrong, or to have a view of the world that misses out something important. But, argues Scott, because the state is powerful, its misperceptions of the world often take physical form, producing well-meaning but clumsy and oppressive modernist schemes that ignore local knowledge and stifle local autonomy.[34] Perhaps our frustrated land registry surveyor decides to write on her clipboard that the local government owns the land; then a few years later the villagers are surprised to find the land being cleared for a palm oil plantation.

One can take the argument even further: that governments can be utterly malevolent, and the worst cases are so catastrophic that they should inform our thinking about how much data any government should have. Wouldn't it have been better if Hitler, Mao and Stalin had understood less about their own societies? Might they have done less harm? And is it reasonable to worry that the more governments know about us, the more they will be tempted to exert control over us?

This argument seems plausible, but I'm not convinced. From communist East Germany to modern-day China, governments interested in mass surveillance and population control have tended to use very different methods to those deployed by independent statistical offices in modern democracies, and to collect very different kinds of data. And history suggests that dictators often have either little interest in the collection of solid statistics, or little ability to collect them.

Consider the disastrous government-induced famine of the late 1950s caused by the Great Leap Forward in communist China, in which people were reduced to eating tree bark, bird droppings and rats. Between 20 and 40 million people died. The catastrophe was made worse by a lack of accurate data about agricultural production. When official statistics began to make the death toll apparent, they were destroyed.[35]

Stalin, similarly, suppressed the publication of the 1937 census of the Soviet Union when it showed that the population was lower than he'd previously announced. This contradiction was an affront in its own right, but it also highlighted the millions of deaths as a result, directly and indirectly, of Stalin's brutality. The penalty for accurately counting the Soviet population? Olimpiy Kvitkin, the statistician in charge, was arrested and shot. Several of his colleagues met the same fate.[36] This is not the act of a totalitarian leader who finds accurate statistical information to be an indispensable tool of oppression.[37]

In Nazi Germany, there was no lack of ambition to use data to support the apparatus of the state. The Reich tried to use punch-card machines, the latest technology, in an effort to track the entire population. But as Adam Tooze argues in *Statistics and the German State*, statistical standards actually fell apart under the Nazis: 'no workable system was ever devised'.[38] The traditions of official statistics – privacy, confidentiality and independence – were so alien to the Nazi project that the system all but collapsed under the political pressure and factional infighting.

All that said, I have a great deal of sympathy with James C. Scott's argument (I discuss Scott's ideas in more detail in my book *Messy*) and some sympathy with Sir John

Cowperthwaite's. States should be humble. Bureaucrats must recognise the limits of their knowledge. There is always a risk that the bird's eye view is so grand and sweeping as to induce delusions of omnipotence.

Sir John's strategy to deny information to the British government seems to have worked for Hong Kong fifty years ago, but Hong Kong was in a very particular situation – a colonial possession of a fading imperial power in which big government was fashionable, and any government intervention would have taken place at a distance of 6000 miles. Those are unusual circumstances.

But the tactic of simply refusing to collect basic statistics could only make sense for a libertarian, laissez-faire regime. And the truth is that very few people seem attracted by that prospect. For better or worse, we want our governments to take action, and if they are to take action they need information. Statistics collected by the state make for better-informed policies – on crime, education, infrastructure and much else.

In poor countries, where official statistical agencies tend to be less well resourced, there is especially wide scope to improve decision-making through better statistics. One example may illustrate the problem. How effective is education in improving literacy? That seems like the kind of question that might usefully help to inform education policy and spending. So researchers at the World Bank looked into statistics collated by UNESCO (the UN Educational, Scientific and Cultural Organization) and found there was an amazingly high correlation between education and literacy: without fail, countries that provided more years of formal education to more people had higher literacy rates. Clearly, education worked! They excitedly published their findings.[39]

Unfortunately, they hadn't read the small print. UNESCO simply hadn't had the resources to collect all the data they wanted to: they had just seventy staff covering 220 countries trying to pull together data in all kinds of areas – adult literacy was just one. (What does literacy even mean in a place such as Papua New Guinea? It has four hundred languages, some of which have no written form.) Inevitably, there would be shortcuts. UNESCO couldn't send teams of people to assess rates of adult literacy themselves, so they looked for a proxy indicator – a best guess in a difficult situation. And they decided that if someone had fewer than five years' formal education, they would be assumed to be illiterate. No wonder the World Bank researchers found such a close correlation between education and literacy.

If organisations like UNESCO had more resources to collect statistics, they would have less need to rely on proxies, and researchers would have greater ability to answer questions such as how well education promotes literacy. Statistical bedrock is so patchy in poor countries that already one dollar in every three hundred that is spent on international aid goes towards funding statistics. There is a case that doubling that might well produce much more value from the remaining $298.[40]

Sir John's comment to Milton Friedman contains an implicit assumption: that government statistics are not just collected by government, but they are collected *for* government. He was unusual in believing that government would do a better job without those statistics, but otherwise that perspective is common. Congress seemed to have the same idea in mind when creating the Congressional Budget Office: the CBO was designed to provide information *to Congress.* The clue is in the name. And the idea goes back a long way. As the future US President James Madison put it in 1790, politicians should be willing to commission accurate statistics, 'in order that they might rest their arguments on facts, instead of assertions and conjectures'.[41]

There is nothing wrong with the idea that government should collect statistics to inform itself. But there is a risk that this view slips into a proprietorial sense of ownership, when politicians believe not only that they should be using statistics to run the country, but that those statistics are none of anyone else's business, and that external scrutiny is a distraction. The facts are no longer the facts – they become the tools of the powerful.

Sir Derek Rayner was a proud proponent of the view that statistics should be managerial tools.[42] Sir Derek had already been a highly successful manager at Marks & Spencer, national treasure of the British high street, before advising the UK government on how to become more efficient. In 1980, Prime Minister Margaret Thatcher asked him to review the way official statistics were collected and published in the UK. Sir Derek was happy to oblige: he saw these numbers as basically a management information system. Those that helped the government run the country could be retained; those that did not could be discarded. And there was no need to make a big fuss about publishing the numbers so that anyone could learn from them, or challenge them.

Sir Derek's view was a mistake. Good statistics don't just serve government planners: they are valuable to a far wider group of people. In the commercial sector, businesses rely on government-collected data to plan their production targets, the location of factories, offices and stores, and other business activities. Data gathered by the Bureau of Labor Statistics, the Census Bureau, the Energy Information Administration and the Bureau of Economic Analysis allow banks, real estate agents, insurance companies, auto manufacturers, construction firms, retailers and many other firms to make plans and to assess their own data against a broader backdrop. The multi-billion-dollar turnover of data-intensive private sector companies such as Bloomberg, Reuters, Zillow, Nielsen and IHS Markit suggests that businesses are willing to pay handsomely for useful statistics; what is less well understood is that these businesses build their statistical edifices on the foundations of government data.[43]

This isn't just about making money: it's about making sure that citizens have access to accurate information about the world in which they live. Government statistical agencies typically make their work available to all, free of charge. Some of that data might be impossible for a private agency to collect, at any price: governments can legally require a response that a private agency could not, as with the case of the census. Other data could be collected but they would be offered only on an expensive subscription basis – private providers can charge tens of thousands of dollars a year for people who want data at their fingertips. Of course some data might be gathered by private firms and given away without charge, but such statistics are often just adverts in the guise of information.

Publicly available statistics can be used to understand and illuminate pressing social issues. To pick just one example, W. E. B. Du Bois – historian, sociologist and civil rights campaigner – led a remarkable data visualisation effort at the end of the nineteenth century as part of the Paris Exposition of 1900.[44] His team produced beautiful, modernist graphs showing the situation of African-Americans in the United States at the time, with data on demographics, wealth, inequality and more besides. Some of them used data that Du Bois and his team had gathered at Atlanta University; but some of the most striking graphics relied on official statistical sources such as the US census. It's just one example of the way in which those who want to understand the world, campaign for change, or both, can turn to official statistics to help them.

With reliable statistics, citizens can hold their governments to account and those governments can make better decisions. If the government decides instead that the statistics belong to politicians, not to citizens, the quality of government decisions will not improve as a result. Neither will the esteem in which government is held.

Sir Derek Rayner's ideas appalled many statisticians. The problem was partly the corrosive message to the British public: 'these numbers aren't for you – they're only for important people'. But even if, like Sir Derek, one believes that statistics really are just for the important people, there's still a good reason to make them publicly available: doing so keeps them honest. As we saw in the previous chapter, public scrutiny is vital. It's what distinguishes science from alchemy. If statistics are published and designed to be accessible to all, they can be analysed and examined by academics, policy wonks and indeed anybody with a bit of time and access to a computer. Errors can be identified and corrected.

As it was, Sir Derek's proposed reforms led to a situation where the definition of unemployment was tweaked more than thirty times in a decade, generally in a way as to lower the headline unemployment rate.[45] That is what happens when statistics are no longer regarded as a public good. And unsurprisingly, people became extremely cynical about the quality of these statistics. 'Phony', as Donald Trump might have said. Of course when official data keep being tweaked for propaganda reasons, trust will rightly evaporate.

The UK's statistical system, now reformed, has spent a quarter of a century trying to recover its reputation. That has taken time and hard work, because trust is easy to throw away and hard to regain. Still, the UK's Office for National

Statistics is more trusted than comparable organisations such as the Bank of England, the courts, the police and the civil service – and vastly more trusted than politicians or the media.[46]

Sir Derek's view – that government-collected statistics mainly exist for the convenience of government administrators, and that citizens have no particular right to see them – has thankfully fallen out of fashion around most established democracies. But one vestige clearly remains, and it was unwittingly highlighted by President Trump on Friday, 1 June 2018 – the day on which the monthly jobs report was to be published.

'Looking forward to seeing the employment numbers at 8:30 this morning,' Trump tweeted, at 7.21 a.m., in an uncanny demonstration of how to wink on Twitter. Markets leaped in expectation of good news. Sixty-nine minutes later the jobs report was released, and – surprise, surprise – the news was indeed good.

Was Mr Trump clairvoyant? No. He had simply been given advanced sight of the job numbers, and decided to tell the world to expect good news.

Official statistics are often both politically and financially sensitive – for example, if the latest numbers on unemployment show that lots of jobs have been created, financial markets will respond in a different way than if the report looks grim. The numbers sometimes shape political arguments, too. For this reason, official statistics are kept confidential as they are being calculated and checked; they are then released at a particular moment, on the dot.

But in some countries, including the US and the UK, certain people get to see certain official statistics in advance. This is called 'pre-release access', and it's a controversial practice. The justification for it is to allow ministers to prepare a response, to answer questions from journalists, and so on. For this reason various political advisers, press officers and the like are often in the list of people given this privileged access. A self-congratulatory review of the practice by the Cabinet Office in the UK noted that press officers thought that ending the practice of pre-release 'would be a disaster . . . The media would simply have their stories without any proper, official comment.' Boo hoo.[47]

It's clear why politicians in power might find it convenient to get advance notice of statistics so they can plan to crow about them if they're good – or if they're bad, to get their story straight or create a distraction. But it's far from clear that this is in the public interest. Why shouldn't everyone, on all sides of the debate, get access to the numbers at the same time, once they're ready?

(There is a compromise position: ministers could receive the statistics thirty minutes in advance and sit alone, without access to a cellphone, to compose a response. Quite apart from being pleasingly like sending powerful people back to sit exams, this is how journalists are sometimes given sensitive official releases. We cope. I was told a story about a Canadian statistician explaining this approach at an international gathering of colleagues. Her Russian counterpart chimed in with a question: 'How does that approach work if the minister wishes to change the statistics?' Exactly.)

There's more at stake here than a sense of fair play. In the UK, where a number of officials and advisers have routinely had pre-release access to the unemployment statistics, market-watchers noticed something strange: key financial market prices such as foreign exchange rates or the price of government bonds would sometimes move sharply not long before the numbers were published. Most of the time this happened, the data would be surprising – either much better or much worse than the market had expected – and the trading would be in the direction that took advantage of the surprise.

Just to check that the market wasn't somehow figuring out the same thing that the statisticians did, forty-five minutes in advance of publication, economist Alexander Kurov made a systematic comparison of the situation in the UK and the situation in Sweden – which is economically quite similar to the UK but which bans pre-release access to official statistics. Swedish politicians and their press officers learn about the numbers at the same time as everyone else – and traders of the Swedish krona, it seems, do not have the same weird powers of clairvoyance as traders of the British pound.[48]

It's impossible to prove, but it seems highly likely that someone with pre-release access was giving the nod to his or her trader friends, allowing insider trading on official data. Who? Well, there were 118 people with pre-release access to the unemployment statistics, which doesn't make it easy to identify a specific culprit. (If you are wondering why it took 118 people to prepare 'proper, official comment' for the media, so am I.)

Mr Trump's tweet probably didn't do much harm in itself: after all, everyone had access to the tweet at the same time. Indeed the President may unwittingly have done some good, by turning the hidden scandal of pre-release access – and the way it is an invitation to corruption, at least when the data go to subtler operators than Mr Trump – into a widely discussed blunder.

Such privileged access facilitates insider trading – but perhaps more important is that it is corrosive of trust in official statistics. The UK press officers, keen to retain the insider perk, protested that if ministers weren't able instantly to offer some polished patter about the data, trust in statistics would be damaged. But the truth is that the countries which are most scrupulous about forbidding pre-release access are also the countries with the strongest public confidence in official data. Those press officers might be surprised at that. I'm not.

Thankfully, the data detectives were at hand to lead the charge. In the UK, the Royal Statistical Society campaigned hard against the practice of letting ministers and other insiders sneak a peek at valuable data before the rest of us. The idea that the government needed to see the numbers so as to compose a press release, said the RSS, 'is pernicious. It skews debate over the figures and perpetuates the impression that ministers control the data.' That seems right to me. In the UK, our levels of trust in official statistics are not as high as in some countries, and not as high as they should be – but they are still far higher than our levels of trust in politicians. I can see why politicians would like to get themselves wrapped up in the release of trusted statistics; it's far from clear why any of the rest of us should want that.

So I'm delighted to report that as of 1 July 2019, the UK decided to emulate the Swedes and end pre-release access to official statistics. Under the new system the only people who will know these numbers before they're published will be the statisticians working on them. Something tells me that trust in official statistics will survive the shock of ministers knowing the facts at the same time as everyone else.

This chapter has been a wholehearted defence of my fellow nerds, the ones who do an essential job in government, sometimes facing indifference from voters, interference from the powerful, and scepticism from all sides.

I wouldn't want to suggest that any nation's official statistical machinery is by definition unimpeachable. We've seen that the official statistics emerging from Argentina and Greece turned out to be deceptive, that the unemployment data emerging from the UK were tweaked every few months throughout the 1980s, and that in Canada statisticians have been forced to resign in protest at the decisions politicians have made. Some statisticians have had to endure death threats made against their families; others openly acknowledge that ministers can change data if they wish. It would be naive to assume that such problems are always exposed and that the truth always triumphs.

Even when official statistics are produced as skilfully and independently as we'd hope, they will never be perfect. Some things we care about are simply hard to measure, such as domestic violence, tax evasion or rough-sleeping. There is, no doubt, plenty of scope for official statisticians to make the data they collect more representative, more relevant, easier to reconcile with everyday experience, and fully transparent. The more they are able to do this, the more they will deserve our trust.

Yet for all their problems and weaknesses, official statistics are still the closest we have to data bedrock. When a country picks and defends a team of skilled, professional and independent statisticians, the facts have a way of making themselves known. When a country's national statistics fall short, an international community of statisticians will complain. When an independent statistician is attacked or threatened by politicians, that same community will rally to his or her defence. Statisticians are capable of greater courage than most of us appreciate. Their independence is not something to take for granted, or to casually undermine.

As citizens, we need to look for that statistical bedrock. If we want to understand the situation a country is in – whether to inform our own decisions, or to hold our government to account – then we will usually start with the statistics and the analysis produced by organisations such as the Office for

National Statistics, Eurostat, Statistics Canada, the Bureau of Labor Statistics and the Congressional Budget Office.

Tough, independent-minded statistical agencies make us all smarter. So: be grateful to Andreas Georgiou, Graciela Bevacqua, and for that matter the late Alice Rivlin. And, if you like, raise a glass to Fanne Foxe.

RULE NINE

Remember that misinformation can be beautiful too

> We are in danger of making the same statistical mistakes that we've always made – only prettier.
>
> —MICHAEL BLASTLAND, co-creator of BBC Radio 4's *More or Less*

Florence Nightingale would have needed no introduction in Victorian Britain: she was the nation's unofficial patron saint, and was the only non-royal woman to appear on English banknotes until 2002. Her legend continues to this day; the four-thousand-bed London hospital constructed in a few days to meet the demands of the pandemic was named Nightingale Hospital.

In Florence Nightingale's own time, the only woman more recognisable would have been Queen Victoria herself. The nation revered Nightingale for her 'feminine' heroics in the Crimean War, pacing the wards of the Scutari barracks hospital in Istanbul. Here's an editorial in *The Times* from

8 February 1855: 'She is a ministering angel without any exaggeration in these hospitals, and as her slender form glides quietly along each corridor, every poor fellow's face softens with gratitude at the sight of her.'

Bleurgh. I'm much more interested in her contribution as a statistician.

Nightingale was the first woman to be made a fellow of the Royal Statistical Society. When her 'slender form' wasn't too busy gliding along the corridors, causing faces to 'soften with gratitude', she was spending her time in Scutari carefully compiling data about disease and death. What she saw in the figures inspired her with a mission to change both the British army and the British nation. Shortly after her return from Crimea, at one of the intellectual dinner parties she often attended, she met William Farr. Farr, thirteen years her senior, had been born to poor parents and lacked Nightingale's fame, front-line experience and political connections. But he was the best statistician in the country, and that was what mattered to her. They became friends and collaborators. One of Nightingale's many biographers, Hugh Small, convincingly argues that the skilful way in which she and Farr wielded the data she'd assembled ended up raising life expectancy in the UK by twenty years and saving millions of lives.[1]

There's a famous remark in a letter between Nightingale and Farr, written in the spring of 1861: 'You complain that your report would be dry. The dryer the better. Statistics should be the dryest of all reading.' It is reported by several biographers as being written by Farr to Nightingale. That makes sense – the fusty middle-aged statistician advising the fiery young advocate to rein in her righteous campaigning impulses. In fact, the biographers are wrong. The letter was written by Nightingale to Farr.* The pair of them were wrestling with the problem of how best to communicate with statistics, and Nightingale was affirming that communications had to be based on hard, dry facts. (In the same letter she wrote: 'We want facts. "Facta, facta facta" is the motto which ought to stand at the head of all statistical work.')[2]

But that didn't mean the communications themselves had to be dry. Nightingale could conjure up an arresting turn of phrase – she argued, for example, that the needlessly high death rates in the army in peacetime were the equivalent of taking 1100 men out on to Salisbury Plain and shooting them.

More pertinent for our purposes, she designed an image that was a landmark in data visualisation. Her 'rose diagram' was arguably the first ever infographic. That makes her perhaps the first person to grasp that busy, influential people would pay far more attention to a vivid diagram than to a table of numbers. In one letter, written on Christmas Day 1857 – less than three years after being beatified in *The Times* – she sketched out a plan to use data visualisation for social change. She declared her plan to have her diagrams glazed, framed and hung on the wall at the Army Medical Board, Horse Guards and War Department. 'This is what

---

* How could multiple experts on Nightingale get this detail wrong? It's unclear who first made the error, but once made, it spread. I've repeated the mistake myself in an article for the *Financial Times*. My first inkling that there was a problem was when I consulted a biography of the far-less-famous William Farr, which says the letter was written to Farr, not by him. I contacted the wonderful archivists at the British Library and discovered that the letter in question is an unsigned draft; the final version is lost. The draft is in the handwriting of Dr John Sutherland, a close collaborator of Nightingale who would often draft for her and may have been taking dictation. It was definitely intended for Farr. Even if it was not dictated by Nightingale it likely reflected her views closely. Professor Lynn McDonald, the editor of a multi-volume collection of Nightingale's work, explained to me, 'She, perhaps, wrote her own version and sent it, but evidently such letter has disappeared. [Sutherland and Nightingale] saw eye to eye – they are her views, AND his.' (Email correspondence, 31 May 2019.)

they do not know and did ought to!', she wrote. She even planned to lobby Queen Victoria, and she knew all too well that beautiful diagrams would be essential. As Nightingale quipped when sending one of her analytical books to the Queen, 'She may look at it because it has pictures.'[3]

It's a cynical, almost contemptuous, thing to write. But it is true. A chart has a special power. Our visual sense is potent, perhaps too potent. The word 'see' is often used as a direct synonym for 'understand' – 'I see what you mean'. Yet sometimes we see but we don't understand; worse, we see, then 'understand' something that isn't true at all. Done well, a picture of data is worth the proverbial thousand words. It is more than persuasive; it shows us things we could not have seen before, revealing patterns amid chaos. However, much depends on the intent of the chart's creator, and the wisdom of the reader.

This chapter, then, will talk about what happens when we try to turn numbers into pictures. We'll see what can go wrong. And by following the story behind Nightingale's famous rose diagram, we'll see how powerful data visualisation can be when used clearly and honestly.

Much of the data visualisation that bombards us today is decoration at best, and distraction or even disinformation at worst. The decorative function is surprisingly common, perhaps because the data visualisation teams of many media organisations are part of the art departments. They are led by people whose skills and experience are not in statistics but in illustration or graphic design.[4] The emphasis is on the visualisation, not on the data. It is, above all, a picture.

The most egregious examples of numbers as decoration are nothing more than the same old number in a large, striking font.

19 – the number of words in the preceding sentence.

I suppose this brightens up a page loaded with text, but it's hardly an insightful use of ink. Also, the correct number is twenty-one. Never let zippy design distract you from the possibility that the underlying numbers simply might be wrong.

Another decorative approach is what we might call 'Big Duck' graphics.[5] The Big Duck itself is a building near New York City, built by a duck farmer in the 1930s to serve as a shop from which he could sell duck eggs and ducks. It may not entirely surprise you to read that the Big Duck closely resembles a 30-foot-long white duck. The architects Denise Scott Brown and Robert Venturi used the term 'ducks' to describe any building that is designed to resemble a relevant product or service, such as a strawberry stand in the shape of a giant strawberry, or Shenzhen airport, which is the shape of an aeroplane.

The graphic guru Edward Tufte borrowed the 'duck' term to describe a similar tendency in graphics: a graph about the NASA budget in the shape of a rocket; a graph about higher education in the shape of a mortar board hat; or, in the example created for *Time* by Nigel Holmes, a graph about the price of diamonds in the shape of a diamond-clad dame, her shapely fishnet-clad legs sketching out the price of a flawless 1-carat stone. Sometimes these visual puns do help people to read and remember the information they frame.[6] But they are often a poor attempt at humour, or a desperate bid to inject interest into data that seem dull. Data visualisation ducks can be more than tasteless: the duckness of the graph can actually obscure – or worse, misrepresent – the underlying information.*

There is a curious historical parallel for this: dazzle camouflage. Dazzle was a defence mechanism for battleships in the First World War, always at risk of being torpedoed by a lurking submarine. The usual 'blending in' method of camouflage wasn't an option for a huge steel vessel that advertised its presence against an ever-changing sea and sky with bow waves and smokestacks. Dazzle camouflage flipped the idea of camouflage on its head. It was an abstract riot of squiggles and harlequin patterns – in fact, it bore enough of a resemblance to Cubist art that Picasso himself impishly tried to claim the credit.[7]

The real inventor of dazzle was Norman Wilkinson, a charismatic artist who joined the Royal Navy reserves at the beginning of the war. He later explained, 'Since it was impossible to paint a ship so that she could not be seen by a submarine, the extreme opposite was the answer – in other

---

* Admittedly, sometimes the underlying data, when plotted in a straightforward fashion, do suggest a picture emerging from the dots on the graph like a Rorschach test. For example, if you plot a chart of unemployment versus inflation in Japan – what economists call the 'Phillips Curve' – you may notice something. As a 2006 economics paper observed, 'Japan's Phillips Curve Looks Like Japan'.

words, to paint her, not for low visibility, but in such a way as to break up her form and thus confuse a submarine officer as the course on which she was heading.'

Because torpedoes took some time to slice through the water to hit their target, the submarine's periscope operator had swiftly to judge a ship's speed and direction before firing the torpedo on an intercept course. Gazing through a tiny scope at a ship in dazzle camouflage, the operator knew he was looking at a ship, but would find it hard to make out any of the clues that mattered for aiming the torpedo accurately. The squiggles looked like bow waves, while harlequin diamonds could easily be confused with the various angled surfaces of a battleship's hull. The result was that the lookout could easily misjudge the ship's speed, the angle of its travel, and the size of and therefore distance to the ship. He might even see two ships rather than one, or mistake the bow for the stern and aim behind the ship rather than ahead of it. Dazzle camouflage was intended to provoke misjudgements.

More than a century later, it isn't hard to see echoes of dazzle camouflage in infographics. From TV to newspapers, websites to social media, we are surrounded by graphical images that grab our attention, pleading to be shared and retweeted, but which also – intentionally or not – mislead, prodding us to a judgement that is often mistaken. At least the periscope operator whose eye was caught by a dazzle ship would have realised that he was looking at something odd, even if he couldn't make sense of it. But many of us who are dazzled by infographics don't suspect a thing.

All that lay far in the future when Florence Nightingale was a girl discovering a passion for data. At the age of nine she was categorising and graphing the plants in her garden. As she grew older, she successfully pleaded with her parents to receive high-quality mathematical tuition; at dinner parties she met the likes of Charles Babbage, a mathematician and designer of a now-famous proto-computer; she was a house-guest of Ada Lovelace, Babbage's collaborator; and she corresponded with the great Belgian statistician Adolphe Quetelet. Quetelet was the person who popularised the idea of taking the 'average' or 'arithmetic mean' of a group, which was a revolutionary way to summarise complex data with a single number. He also pioneered the idea that statistics could be used not just to analyse astronomical observations or the behaviour of gases, but social, psychological and medical questions such as the prevalence of suicide, obesity or crime. Babbage and Quetelet were later to be founders of the Royal Statistical Society; Nightingale, as I have mentioned, became its first female fellow.

By her thirties, Florence Nightingale was steeped in this world of pioneering mathematicians – but her job was as nursing superintendent at a small hospital in London's Harley Street, where she not only sorted out the book-keeping and hospital infrastructure, but sent surveys to other hospitals all around Europe, asking them about their administrative practices and tabulating the results.

It was at this time, late in 1854, that she was persuaded by the Secretary of State for War, her old friend Sidney Herbert, to lead a delegation of nurses to Istanbul to tend to wounded British soldiers from the Crimean War. The war was a bitter struggle between the Russian Empire and several other great European powers, including Britain. Nightingale's presence with the British army, which was unprecedented for a woman, was designed to pacify a public incensed by what they were reading about the dreadful conditions in the hospitals there. Reports in *The Times* turned the Crimean War into a long-running narrative of disaster with many familiar characters. By the end of it all, Florence Nightingale was perhaps the only figure to retain public support; the generals and the rest had been discredited by the catastrophe.

The Barrack Hospital in Scutari, Istanbul, was a death trap. Hundreds of soldiers from the Crimean front were succumbing to typhus, cholera and dysentery as they tried to recover from their wounds in cramped conditions next to the sewers. Nightingale arrived to find rats and fleas everywhere she looked. Basics such as beds and blankets were missing, as were food to cook, pots to cook it in, and bowls to eat it from. All this outraged public opinion when it was reported in *The Times*, and Nightingale herself was quick to use the newspaper to raise funds from readers – and to pressure an ill-organised British army to get its act together.

Less of a *cause célèbre* was that the hospital record-keeping was as badly organised as anything else. There were no standardised medical records and no consistent reporting between the various British army hospitals. This may seem a relatively trivial matter, but Nightingale knew it was a big problem. Without good statistics it was impossible to understand why so many soldiers were dying, or to find a way to improve conditions. Even the dead were going uncounted, buried without their deaths being recorded. Nightingale saw all this more closely than anyone. She even took on the duty of writing to the family of each dead soldier. But she wanted the bird's eye view as well as personal experience, understanding that certain truths can only be perceived through the statistical lens. She tried to standardise and make sense of the hospital data.

Long after the war was over, Nightingale was still pressing to improve the standard of medical statistics. Some of this work, in partnership with Farr, was magnificently unglamorous. For example, they tried to standardise the description of different illnesses and causes of death, with Farr leading on the technical side while Nightingale campaigned for his ideas to be adopted. She wrote to the International Statistical Congress in 1860 to argue that hospitals should make use of Farr's methods by collecting statistics according to a uniform standard. This wasn't mere fussiness: standardising the statistics meant that different hospitals could be compared and could learn from each other. It is this sort of statistical foundation-building that many of us overlook – but as we've seen many times during this book, without well-defined standards for statistical record-keeping, nothing adds up. Numbers can easily confuse us when they are unmoored from a clear definition.

Florence Nightingale may have been a savvy campaigner, but her campaigns were built on the most solid of foundations.

The most straightforward problem with a clever decorative idea is that the basic data may not be solid. The visualisation then simply hides that fact – the shimmering icing over a mouldering statistical cake.

One educational example is *Debtris*, an unforgettable animation produced several years ago by David McCandless, author of *Information is Beautiful*.[8] It shows large blocks falling slowly against an eight-bit soundtrack in homage to the addictive computer game Tetris. Their size indicates their dollar value. '$60bn: estimated cost of Iraq war in 2003' is followed by '$3000bn: estimated total cost of Iraq war', and then Walmart's revenue, the UN's budget, the cost of the financial crisis, and much else. As decoration this is wonderful stuff: the graph looks great, the music endlessly loops in your head, and the slow revealing of the different comparisons leave you gasping with surprise, laughter and anger.

But the same elements that make *Debtris* such a delight to watch also make it much harder to spot the underlying problems. Statistical apples are compared with statistical oranges throughout. Stocks are compared with flows. That's the equivalent of comparing the total cost of buying a house with the annual cost of renting one; it's not a trivial confusion. Net measures are put alongside gross ones – the equivalent of comparing a firm's profit with its turnover.

The shocking difference between the before-and-after comparisons of the cost of the war in Iraq turns out to be based on an unfair comparison. (Admittedly, a fair comparison might *also* show a shocking difference.) The pre-war number is a narrow estimate: the cost to the US military budget. The post-war number is very broad, including a figure for the cost of lost life, the cost of high oil prices, and a huge sum for the cost of macroeconomic instability, putting part of the blame for the 2008 financial crisis on the war. That broad estimate of cost is not unreasonable, but what *is* unreasonable is to put it alongside a very different kind of estimate without comment. What seems to be a pure before-and-after contrast is actually narrow-and-before versus broad-and-after, measuring a different thing at a different time. Nobody looking at the *Debtris* animation would realise that.

Debtris was published in 2010, and quickly became my favourite cautionary example – the visualisation is so good but the data are so bad. A couple of years later I was introduced to David McCandless at a conference. I felt a bit awkward. I'd been moaning about his work while he wasn't in the room, but I'd never done him the courtesy of sending him an email with my comments. But perhaps he hadn't noticed them? I felt compelled to confess.

'I should probably say, David, that I have a concern about your *Debtris* animation.'

'I know you do,' he replied.

I squirmed. But to his credit, his more recent work is similarly striking, while also being more careful about the underlying data. For example, a visualisation in a similar spirit – 'The Billion-Pound O'Gram' – still mixes stocks and flows, but it is much more transparent about doing so.[9] Further to McCandless's defence, the only reason I could find out that the data behind his *Debtris* animation were patchy and inconsistent is that he fully referenced it. Many don't.

So information is beautiful – but misinformation can be beautiful, too. And producing beautiful misinformation is becoming easier than ever.

Graphics once required a great deal of time and trouble to produce and then reproduce. Even something as simple as a graphic with straight lines, precise edges and colour would have demanded expert draughtsmanship and expensive printing methods. It's telling that Edward Tufte, in a 1983 book, devotes some attention to deploring the use of diagonally shaded black-and-white patterns because they can produce an unsettling optical illusion of a flicker. 'This moiré vibration [is] probably the most common form of graphical clutter,' he complains. It may have been common then; it is unheard of today. We would now invariably use colour rather than diagonal shading.

No draughtsmanship is required these days. A variety of powerful software tools can swiftly turn numbers into pictures. But any powerful tool should be used with care, and the very speed of the process means that impressive-seeming graphics can be created without any serious thought about the underlying data or how best to describe them.

The ease of creating pretty graphics is exceeded only by the ease of sharing them. A quick 'like' on Facebook or a retweet on Twitter will speed the image on. Ideas that are best expressed in words or numbers are turned into graphics anyway, because that's what spreads on social media. Unfortunately the selection mechanism is often some combination of beauty and shock value, rather than pertinence and accuracy.

Consider the experience of Brian Brettschneider, a climate

scientist with a fondness for gorgeous maps. He celebrated Thanksgiving in 2018 by producing a map showing 'The Favorite Thanksgiving Pie by Region', including coconut cream pie for the mid-west, sweet potato pie for the west coast, and key lime pie for the south. As a Brit, I don't know much about Thanksgiving, and my favourite pie is a cold pork pie, but I'm told that the map seemed wrong to American eyes. No pumpkin pie? No apple pie? The map – and the outrage – went viral on Twitter. Senator Ted Cruz, a prominent Republican politician, didn't like the suggestion that Texans favoured key lime pie: '#FakeNews', he tweeted.

And he was right. Brettschneider had made it all up. He was joking; the map was a parody of all the other bad maps that go viral on the internet. After more than a million people had seen the tweet, however, Brettschneider started to become uneasy. Did people even know he was joking? We don't know who got the joke, who shared the map in mild outrage, and who believed it was solid fact. But we can be fairly sure that the use of a vivid graphic gave it its viral power. 'We tend to place very high value in maps as holders of accurate information,' writes Brettschneider. 'If it's in a map, it must be true, right? If I had tweeted a joke list of favorite pies by region, it would be very quickly ignored. Since it was in map form, it had an air of authenticity.'[10]

Quite so. My only difference with Brettschneider is that I don't think the problem is limited to maps. Any vivid graphic has the potential to go viral, whether true, false, or a bit of both. This book started with a warning that we should notice our emotional response to the factual claims around us. Just so: pictures engage the imagination and the emotion, and are easily shared before we have time to think a little harder. If we don't, we're allowing ourselves to be dazzled.

*

The situation in the Scutari hospital was catastrophic. Florence Nightingale was later to write, 'To inexperienced eyes the Scutari buildings were magnificent. To ours, in their first state, they were truly whited sepulchres, pest houses.'[11] But why, exactly, were so many soldiers dying?

Poor hygiene is the obvious explanation from a modern perspective: germs were being transmitted freely in the filthy, vermin-ridden conditions. But the idea that diseases might be transmitted by microbes, and fought by using antiseptics and keeping things clean, was in its infancy. Very few doctors would even have heard of it as speculation, let alone believed it. Nightingale was no different; she thought instead that the high death toll in Scutari was due to lack of food and supplies, a problem she sought to remedy with her high-profile fundraising and campaigning through *The Times*.

Nevertheless, she also requested a team to help clean up the hospital, and in the spring of 1855 this 'sanitary commission' arrived from the UK, whitewashed the walls, carted away filth and dead animals, and flushed out the sewers. The main hope was to make the hospital less unpleasant, but the immediate effect was to cut the death rate almost immediately from more than 50 per cent to 20 per cent.

Florence Nightingale wanted to understand what had happened, and why. And like Richard Doll and Austin Bradford Hill, she believed she could work out the truth if she examined the data with sufficient care. Her scrupulous record-keeping made the dramatic improvement after the sanitary commission's work very clear indeed.

When Nightingale returned from the war, Queen Victoria summoned her for a royal audience. Nightingale persuaded Victoria to support a Royal Commission investigating the health of the army. She also recommended that the commission include William Farr, though Farr's low-born status meant that he was not treated well by the establishment: he was eventually retained only as an unpaid consultant to the commission.

Nightingale and Farr concluded that poor sanitation had caused many of the deaths in the Crimean War hospitals, and that most military and medical professionals had failed to learn this lesson. The problem was much bigger than one war: it was an ongoing public health disaster in barracks, civilian hospitals and beyond. The pair began to campaign for better public health measures, tighter laws on hygiene in rented properties, and improvements to sanitation in barracks and hospitals across the country.*

Nightingale may have been the most famous nurse in the country, but she was a woman in a man's world, and had to convince the country's medical and military establishments, led by England's chief medical officer, John Simon, that they had been doing things wrong all their lives. Dr Simon wrote in 1858 that deaths from contagious diseases were 'practically speaking, unavoidable' – that there was nothing to be done to prevent future deaths. Nightingale set herself the task of proving him wrong.

William Farr's daughter, Mary, described eavesdropping on an early conversation between her father and Florence Nightingale. Mary recalled Farr giving Nightingale a warning about speaking out against the establishment. '"Well, if you do it, you will make yourself enemies," and she drew herself up and answered, "After what I've seen, I can fire my own guns."'[12]

---

* *The Big Issue* magazine put Florence Nightingale on the cover in March 2020. 'Hail the Hand-Washing Queen: How Florence Nightingale Is Helping Us Fight Coronavirus', it trumpeted. But it takes more than hand-washing to power a revolution in public health: it takes statistical detective work. The virus reminded us to wash our hands, but more important, it taught us that fighting an epidemic requires information, as quick and as complete as possible. Florence Nightingale understood that nearly two hundred years ago. I'd rather remember her, not as a hand-washing queen, but as a data detective.

Nightingale wrote to her friend the Secretary of State for War, Sidney Herbert, 'Whenever I am infuriated, I revenge myself with a new diagram.'[13] Statistics had been the telescope through which she perceived the truth; now she needed a diagram that would compel everyone else to look at the truth, too.

'A good chart isn't an illustration but a visual argument,' declares Alberto Cairo near the beginning of his book, *How Charts Lie*.[14] As the title of the book implies, Cairo has some concerns. If a good chart is a visual argument, a bad chart may be a confusing mess – or it may *also* be a visual argument, but a deceptive and seductive one. Either way, by organising and presenting the data we are inviting people to draw certain conclusions. And just as a verbal argument can be logical or emotional, sharp or woolly, clear or baffling, honest or misleading, so too can the argument made by a chart.

I should note here that not all good charts are visual arguments. Some data visualisation is not intended to be persuasive, but exploratory. If you're handling a complex dataset, you'll learn a lot by turning it into a few different graphs to see what they show. Trends and patterns will often leap out immediately if plotted in the right way. For example, visualisation expert Robert Kosara suggests plotting linear data on a spiral. If there's a periodic pattern to the data – say, repeating every seven days or every three months – that may be concealed by other fluctuations in a conventional plot but will leap out in a spiral plot.

Similarly, certain kinds of problem make themselves known immediately when the data are turned into pictures. Imagine a dataset with the height and weight of tens of thousands of hospital patients. Some of them are 50 or 60 feet tall! That must be a typo. Hundreds of them have a weight of zero. That would be because a nurse or doctor was filling in an electronic form, didn't take a weight measurement, so just hit 'enter' and moved on to the next box. These problems won't be apparent if you ask your computer to calculate an average or a standard deviation, or if you scan columns of data manually. If you look at a picture of the data, however, you'll see the problem in a second.

But let's assume you've explored your numbers, and now you want to turn them into a visual argument. The standard advice for management consultants and academic researchers presenting a graph is to include a title or caption that calls attention to the key features of the data, and draws a conclusion.[15]

Say it with Charts, the bible of management consultants, makes this process very clear. First, says author Gene Zelazny, decide what you want to say with a graph. Once you've decided what you want to say, that suggests a particular kind of comparison. That, in turn, suggests a particular choice of graph – such as a scatter plot, a line graph, a stacked bar chart or a pie chart.* Finally, underline your message by sticking it in the graph title. Don't just write 'Number of contracts, January–August'. Write something like 'The number of contracts has increased' or perhaps 'The number of contracts has been fluctuating', depending on whether you'd like to call attention to the upward trend or to the variations around that trend. Zelazny's vision is one in which the management consultant tells people what to think. Both the graphs and the annotation are chosen to support that message.

I realise there's something unsettling about the way that this process starts with the conclusion and then figures out how to package the data to support that conclusion. But let's be fair: a lot of communication works in this way. Newspaper articles begin with a headline; the rest of the text is explanation. Even a scientific paper begins with an abstract that serves a similar

* Joking. Don't use a pie chart.

purpose to a newspaper headline: it tells you what happened and what it means. A good journalist doesn't begin reporting with the conclusion in mind; a good scientist doesn't decide on the results before the experiment has been run. (I can't vouch for what a good management consultant does.) But once both journalists and scientists have discovered something of interest, they want to give their audiences some pointers as to what it is. The same is true for chart designers.

Edward Tufte, the influential information designer, admires graphics that are dense and complex with a minimum of decoration or annotation. The introduction to one of his books, *Envisioning Information*, sternly warns readers, 'The illustrations repay careful study. They are treasures, complex and witty, rich with meaning.' Look hard. Think. Pay attention at the back of the class. For Tufte, the ideal graphic invites the reader to sit down with a cup of coffee and really pore over the details. 'Emaciated data-thin designs', he warns, 'provoke suspicions – and rightfully so – about the quality of measurement and analysis.'[16]

He may be right – although as we should know by now, the data-density of the graph is no guarantee that the data themselves are reliable: a graph which presents a few data points in a light-hearted fashion may be unimpeachable, while an intricate graphic may be saturated with bad data.

Even if the numbers are solid, a graphic detailed enough to demand a coffee may be persuasive without also being informative. An impressive example is the *New Yorker* website's 2013 presentation of data about inequality. The infographic, designed by Larry Buchanan, evokes the iconic New York City subway map. Viewers can click on different subway lines and see how median income varies along each line. This is an evocative data visualisation 'duck': the graphs of rising and falling income resemble subway routes, and they carefully copy the distinctive design elements of the New York subway map and signage.[17]

What makes the infographic persuasive is that it invites us to make a natural comparison and immediately imagine the people behind it: we observe incomes varying along a chosen line at it moves through different neighbourhoods, we grasp the vast inequality encompassed in a brief subway ride, and we picture the characters involved, rubbing shoulders in the subway car. The rich and poor are so close together, so similar in some ways and yet so different. The infographic carries a real emotional punch.

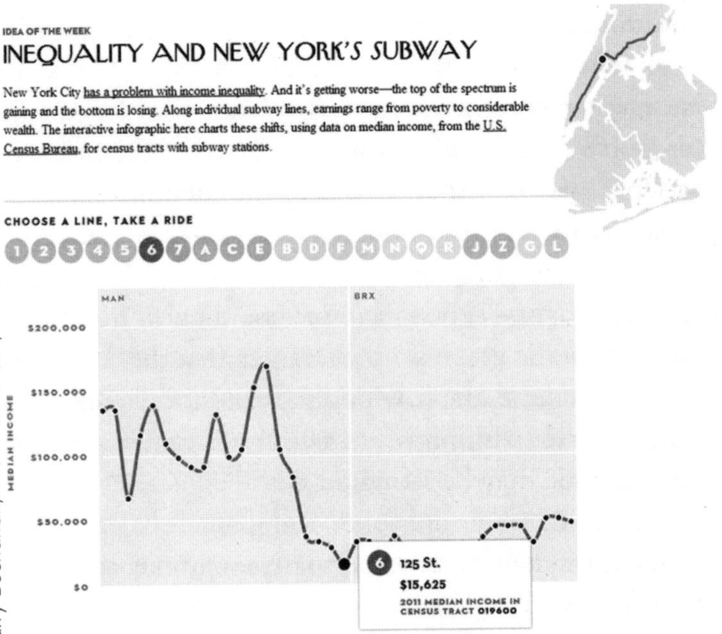

IDEA OF THE WEEK

INEQUALITY AND NEW YORK'S SUBWAY

New York City has a problem with income inequality. And it's getting worse—the top of the spectrum is gaining and the bottom is losing. Along individual subway lines, earnings range from poverty to considerable wealth. The interactive infographic here charts these shifts, using data on median income, from the U.S. Census Bureau, for census tracts with subway stations.

CHOOSE A LINE, TAKE A RIDE

MAN BRX

MEDIAN INCOME

$200,000 $150,000 $100,000 $50,000 $0

6 125 St.
$15,625
2011 MEDIAN INCOME IN CENSUS TRACT 019600

Larry Buchanan, *The New Yorker*, 2013

But is it *informative*? Not so much. As we click around, it is surprisingly hard to learn anything that we didn't already know. It's difficult to compare one subway line with another or to spot any but the most obvious patterns.

This becomes clear when we read the brief article accompanying the infographic, which is full of facts that cannot easily be discovered from the graphic itself. The highest median household income of a subway-endowed census tract in New York City was $205,192. The lowest was $12,288. The article also tells us the subway lines with the largest and the smallest income ranges, and the largest gap between any two stations, although quite why any of this information is useful is unclear. The blog post notes that income inequality in Manhattan is similar to inequality in Lesotho or Namibia. Is that bad? It sounds bad. If you happened to carry around a list of the income inequality recorded in every country on the planet, you'd realise that it *was* bad. But do you? The goal of the graphic is not to convey information but to stir feelings. If the article compared income inequality in New York with other global cities such as London and Tokyo, and other US cities such as Chicago and Los Angeles, we might actually learn something worth knowing.

The result is gorgeous but far less informative than a map would have been. It is a piece of persuasive art pretending to be a piece of statistical analysis. We've been powerfully reminded of something we already believed. We are more passionate, more engaged, but are we truly any more informed?

There's nothing wrong with a polemic – I write them myself, occasionally – but we should be honest with ourselves about what's going on.

Another example is a graph by Simon Scarr, a senior designer at Thomson-Reuters. The graph depicts deaths in Iraq in each month between 2003 and 2011. It's an inverted bar graph: the larger the number of deaths that month, the longer the bar hangs down. Scarr coloured his bars red, meaning the entire graph looks like blood running down from some awful gash at the top of the page. In case the message was ambiguous, the chart is titled 'Iraq's Bloody Toll'. If Larry Buchanan's subway-inequality graph tugs at your heart-strings, Scarr's graph rips your heart right out of your chest. Not for nothing did it win a design award.[18] And unlike the subway diagram, Scarr's graph does give you the relevant information: it is persuasive *and* informative.

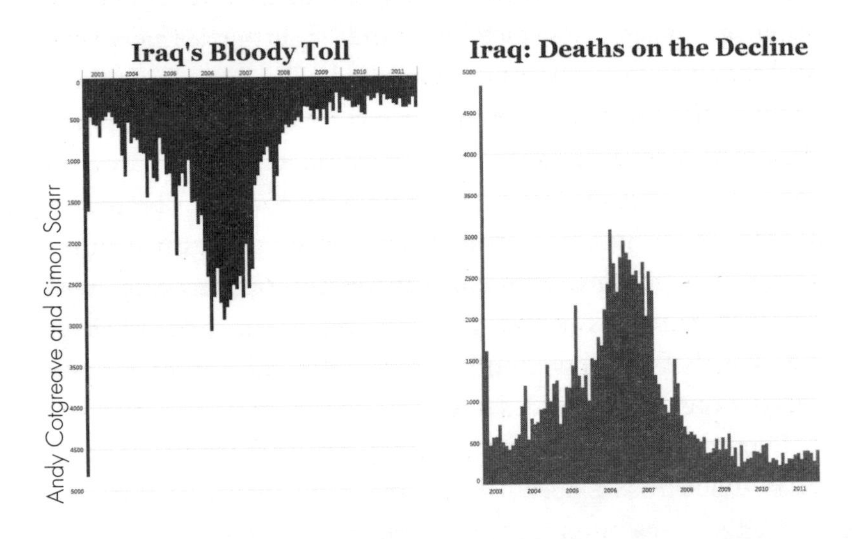

But when Andy Cotgreave, a data visualisation expert, saw Scarr's graph, he tried a little experiment. First, he re-coloured the graph, showing the same bars in a cool corporate blue-grey. Then he turned it upside down. Finally, he changed the title from 'Iraq's Bloody Toll' to 'Iraq: Deaths on the Decline'. The change in the emotional impact is bracing. Scarr's graph screamed raw outrage. Cotgreave's is sober, almost soothing. Which is the better graph? It depends on the message. Scarr's graph wails, 'Oh, the humanity!' Cotgreave's graph calmly states, 'The worst is behind us'. Both messages were fair. It's a reminder that the simplest choices of colour and alignment can change the tone of a chart, and how people will perceive that chart, just as your tone of voice can dramatically alter how your words will be received.[19]

How could a low-born statistician, William Farr, and a mere female, Florence Nightingale, win over the stubborn doctors and soldiers of the Victorian establishment?

First, they had to make sure their data were absolutely watertight. Facta, facta facta! Farr and Nightingale knew very well that their work would be pounced upon by their political enemies. In one telling exchange, Nightingale wrote to Farr telling him to prepare for an attack on his latest statistical analysis. His response showed the confidence he had in the quality of the work: 'Let us wait, & keep our powder dry. We are not going to fire in the air – like people frightened out of their wits. Let them point out our "mistakes", and if they are mistakes – we will admit them freely: but shake our foundations – or blow down our walls – the fellows cannot.'[20]

Then they had to present their findings. Nightingale circulated her 'rose diagram' in 1858, and published it early in 1859. That was just a few years after her time at Scutari hospital – and a matter of months after Dr John Simon's assertion that contagious diseases were practically unavoidable. The rose diagram is a brilliant visual argument. I've seen one of the original prints up close, in the library of the Royal Statistical Society. It's breathtaking, and alarming, a beautiful array of coloured wedges showing deaths from infectious diseases before and after the sanitary improvements at Scutari.

If you wanted to be unkind about the diagram, you could say that it was a pie chart on steroids. Technically, it's a 'polar area diagram', quite possibly the first such diagram ever created. What it *isn't* is a dry presentation of statistical truth. It tells a story.

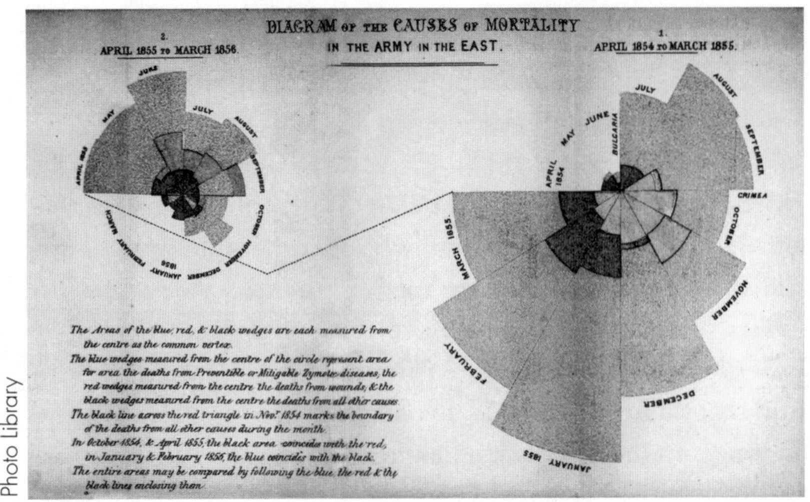

To see just how powerful a piece of visual rhetoric it is, consider the alternative presentation as a bar chart (the example below is based on a graph by Nightingale's biographer Hugh Small, using William Farr's data).

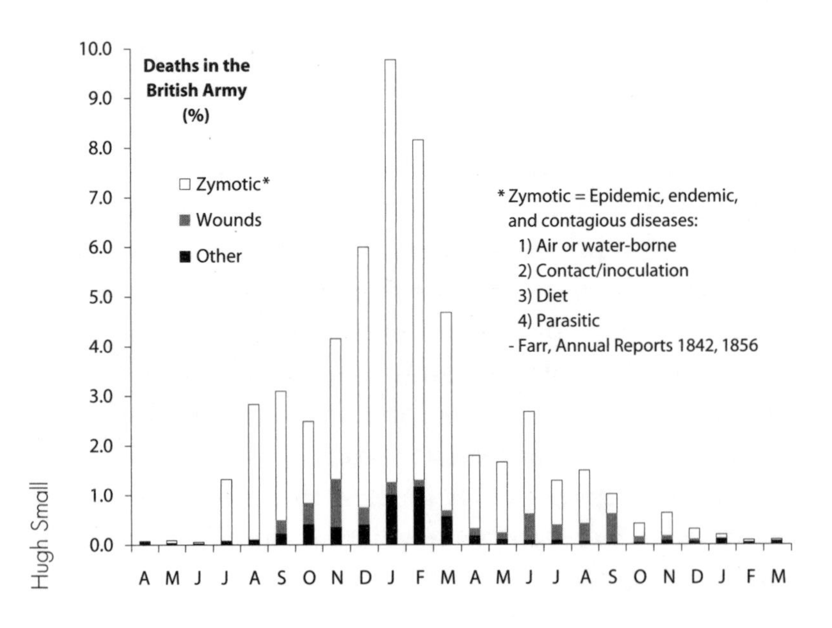

At first glance, Small's bar chart is far clearer and easier to follow. But it draws the viewer towards the wrong conclusion. The bar chart focuses attention on the dreadful death toll of January and February 1855, which might lead one to wonder if these deaths were basically caused by a bitter winter, and spring brought relief. It also makes the decline in deaths look dramatic but smooth – a process rather than a sharp change.

The polar area diagram, in contrast, divides the death toll into two periods – before the sanitary improvements, and after. In doing so, it visually creates a sharp break that is less than clear in the raw data. Because the polar area diagram plots deaths in proportion to the area of a wedge, rather than the height of a bar, it also slightly obscures just how awful January and February 1855 were, instead lumping them together with the grim bulk of 'before the sanitary commission'.

Nightingale wanted to make the importance of improved sanitation leap off the page, convincing the viewer that the Scutari experience could be repeated in hospitals, barracks and even private dwellings across the British Empire. She created the powerful 'before and after' structure of the diagram to strengthen that argument.

Is this dazzle camouflage? Perhaps. I'm inclined to say it isn't, if only because the data are rock solid and there in plain sight. Unlike *Debtris*, it doesn't rely on patchy statistics and unhelpful comparisons; unlike the subway-inequality diagram it isn't all sizzle and no steak. It's more like 'Iraq's Bloody Toll', but it is far more subtle in the way it invites readers to draw their conclusions. Few discussions of the rose diagram highlight just how clever it is at directing the reader towards one interpretation of the data, and not another. Thankfully the idea was both true and important; the visual rhetoric helped people to reach a conclusion that happened to be correct.

Nightingale explained to Sidney Herbert that the diagram 'is to affect thro' the Eyes what we may fail to convey to the brains of the public through their word-proof eyes'. To get her diagram in front of as many eyes as possible, Nightingale asked the radical writer Harriet Martineau to produce a moving book about the Crimean War and the suffering of British soldiers there. Martineau had read Nightingale's reports and praised them as 'One of the most remarkable political or social productions ever seen.' Nightingale included her polar area diagram as a fold-out frontispiece in Martineau's book. It didn't get read by as many soldiers as it might have done – the army banned it from military libraries and barracks[21] – but Nightingale had a more particular audience in mind for her diagram, as she told Herbert:

> None but scientific men even look into the appendices of a Report, and this is for the vulgar public ... Now, who is the vulgar public who is to have it? (1) The queen (2) Prince Albert ... (7) all the crown heads in Europe, through the ambassadors or ministers of each (8) all the commanding officers in the army (9) all the regimental surgeons and medical officers ... (10) the chief sanitarians in both houses [of Parliament] (11) all the newspapers, reviews and magazines.

The senior doctors who had argued that there was nothing to be done gradually came round to Nightingale's argument for better sanitation. In the 1870s, Parliament passed several public health acts. Death rates in the UK began to fall, and life expectancy to rise.

What makes Florence Nightingale's story so striking is that she was able to see that statistics could be tools and weapons at the same time. She appreciated the importance of solid foundations such as the tedious tasks of standardising definitions and getting everyone to fill in the right forms, and of producing 'the dryest of all' analyses, impervious to attack from the critics. But she also understood the need to give the data a makeover, presenting them in the most persuasive light. She produced a picture with enough power to change the world.

Florence Nightingale was on the right side of history, but many of the people who misuse catchy graphics are not. For those of us on the receiving end of beautiful visualisations, everything we've learned so far in this book applies.

First – and most important, since the visual sense can be so visceral – check your emotional response. Pause for a moment to notice how the graph makes you feel: triumphant, defensive, angry, celebratory? Take that feeling into account.

Second, check that you understand the basics behind the graph. What do the axes actually mean? Do you understand what is being measured or counted? Do you have the context to understand, or is the graph showing just a few data points? If the graph reflects complex analysis or the results of an experiment, do you understand what is being done? If you're not in a position to evaluate that personally, do you trust those who were? (Or have you, perhaps, sought a second opinion?)

When you look at data visualisations, you'll do much better if you recognise that someone may well be trying to persuade you of something. There is nothing wrong with artfully persuasive graphs, any more than with artfully persuasive words. And there is nothing wrong with being persuaded, and changing your mind. That's our next subject.

RULE TEN

Keep an open mind

A man with a conviction is a hard man to change. Tell him you disagree and he turns away. Show him facts or figures and he questions your sources. Appeal to logic and he fails to see your point.

—LEON FESTINGER, HENRY RIECKEN and STANLEY SCHACHTER, *When Prophecy Fails*[1]

Irving Fisher was one of the greatest economists who ever lived.[2]

'Anywhere from a decade to two generations ahead of his time,' opined the first Nobel laureate economist Ragnar Frisch, in the late 1940s, more than half a century after Fisher's genius first lit up his subject. Paul Samuelson, who won the Nobel Memorial Prize the year after Frisch, said that Irving Fisher's 1891 PhD thesis 'was the greatest doctoral dissertation in economics ever written'.

That's what Fisher's peers thought. The public loved him too. A hundred years ago, Irving Fisher was the most famous economist on the planet. Yet Fisher is remembered now only by economists with a sense of history. He's no longer a household name like Milton Friedman, Adam Smith or John Maynard Keynes, his younger contemporary. That is because something awful happened to Irving Fisher, and to his reputation – something with a lesson for us all.

Fisher's downfall certainly wasn't through lack of ambition. 'How much there is I want to do!' he wrote to an old school friend while studying at Yale. 'I always feel that I haven't time to accomplish what I wish. I want to read much. I want to write a great deal. I want to make money.'

It was understandable that money was important to Fisher. His father had died of tuberculosis the very week that Irving arrived at Yale. Fisher's drive and intellect kept him afloat: he won prizes in Greek and Latin, for algebra and mathematics, for public speaking (finishing second to a future US Secretary of State), and was both the class valedictorian and a member of the rowing crew. Yet amid all these achievements, the young man needed to scramble for funds throughout his studies; he understood what it was to struggle financially while surrounded by wealth.

At the age of twenty-six, however, Fisher found himself with a small fortune at his disposal. He married a childhood playmate, Margaret Hazard, who was the daughter of a wealthy industrialist. Irving and Margaret's wedding in 1893 was sumptuous enough to be covered by the *New York Times*, with two thousand invited guests, three ministers, an extravagant lunch and a 60lb wedding cake. They commenced a fourteen-month European honeymoon and returned to a brand-new mansion at 460 Prospect Street, New Haven. It had been built in their absence as a wedding present from Margaret's father and was furnished with a library, a music room and spacious offices.

There are three things you should know about Irving Fisher.

The first is that he was a health fanatic. This was

understandable. Tuberculosis had killed the young man's father; fifteen years later, the disease nearly killed him, too. No wonder he adopted a fastidious health regime: he abstained from alcohol, tobacco, meat, tea, coffee and chocolate. One dinner guest enjoyed his hospitality while noting his quirkiness: 'While I ate right through my succession of delicious courses, he dined on a vegetable and a raw egg.'[3]

This wasn't just a personal matter: he was an evangelist for health and nutrition. He founded the 'Life Extension Institute' and persuaded William Taft, who'd just stepped down as President, to be its chairman. (It may seem an ironic choice: Taft was obese, the heaviest man ever to be President. Taft's weight problem, did, however, prompt his interest in diet and exercise.) In 1915, when he was nearly fifty years old, he published a book titled *How to Live: Rules for Healthful Living Based on Modern Science.* (How to live! Now that's real ambition.) It was a huge bestseller, and it's hilarious from a modern perspective. '[I advocate a] sun-bath ... common sense must dictate its intensity and duration' ... 'it is important [to] practice thorough mastication ... chewing to the point of natural, involuntary swallowing'. He even adds a discussion of the correct angle between the feet while walking – 'about seven or eight degrees of out-toeing in each foot'.[4]

And there's a short section on eugenics. It hasn't aged well.

But while it's easy to laugh at the book, *How to Live* is in many ways as far ahead of its time as Fisher's economic analysis. Fisher applied scientific thinking to the question of well-being. He described detailed exercises, preached mindfulness, and at a time when the majority of doctors were smokers, correctly warned that tobacco causes cancer.

That is the second thing you need to know about Irving Fisher: he believed in the power of rational, numerical analysis, in economics and elsewhere. He calculated the net economic cost of tuberculosis. He conducted experimental investigations of vegetarianism and even of thorough mastication, which he found to increase endurance. (A 1917 advertisement for the breakfast cereal Grape Nuts included an endorsement from Professor Fisher.) At one point in *How to Live*, he even pauses to inform the reader that 'in the modern study of scientific clothing there is a new unit, the "clo". This is a technical unit for measuring the "warming power" of clothing.'

It is arguable that his love of numbers occasionally led him astray. For example, when Fisher quantified the benefits of prohibition, he exuberantly generalised from a small study that a stiff drink on an empty stomach made workers 2 per cent less efficient. Fisher calculated that prohibition would add $6 billion to America's economy – which at the time was an absolutely enormous gain. We saw in the first chapter that Abraham Bredius's art expertise had highlighted reasons to believe that Han van Meegeren's rotten forgery was truly a Vermeer. Similarly, Fisher's statistical expertise allowed him to produce grand calculations about prohibition on a shaky foundation stone. His strong feelings about the evils of alcohol were undermining the rigour of his statistical reasoning.[5]

There's also the money – that's the third thing you need to know. Irving Fisher was rich, and not just because of his wife's inheritance. Making money was a matter of pride for Fisher; he didn't want to be dependent on his wife. There were the book royalties from *How to Live.* There were his inventions, most notably a way of organising business cards that was the forerunner of the Rolodex. He sold that invention to a stationery company for $660,000 in cash – many millions of dollars in today's terms – a seat on its board and a bundle of stocks.

Fisher turned his academic research into a major business operation called the 'Index Number Institute'. It sold data, forecasts and analysis as a syndicated package, 'Irving Fisher's Business Page', to newspapers across the United States. Forecasting was a natural extension of the data and analysis. After all, if we want to make the world add up, it's not always because the intellectual joy of understanding is an end in itself. Sometimes we're interested in sizing up the current situation as a means to anticipate, and perhaps profit from, what will happen next.

With such a platform, Fisher was able to evangelise about his approach to investment – which, broadly speaking, was to bet on American growth by buying shares in the new industrial corporations using borrowed money. Such borrowing is often called leverage, since it magnifies both profits and losses.

But during the 1920s, stock market investors had few losses to worry about. Share prices were soaring. Anyone who had made leveraged bets on that growth had every reason to feel clever. Fisher wrote to his old childhood friend to inform him that his ambition had been fulfilled. 'We are all making a lot of money!'

In the summer of 1929, Irving Fisher – bestselling author, inventor, friend of presidents, entrepreneur, health campaigner, syndicated columnist, statistical pioneer, the greatest academic economist of his generation, and a millionaire many times over – was able to boast to his son that a renovation of the family mansion had been paid for not by Hazard family money but by Irving Fisher himself.

That achievement mattered to him. Fisher's own father hadn't lived to see his seventeen-year-old boy grow into one of the most respected figures of the age; as Irving and his son watched a mansion reshaped before them, he could, perhaps, be forgiven his pride. But he was standing on the brink of a financial precipice.

*

The stock market cracked in the autumn of 1929. The Dow Jones Industrial Average fell by more than a third between the beginning of September and the end of November. But it wasn't the great Wall Street crash that did for Irving Fisher – at least, not immediately. The crash, of course, was a cataclysmic financial event, one far more severe even than the banking crisis of 2008. The Great Depression that followed was the greatest peacetime economic calamity to befall the western world. Fisher was more exposed than many, since he had made his investments with leverage, magnifying both losses and gains.

But it took more than a leveraged bet on a financial bubble to ruin Fisher. It took stubbornness. The crash had its dramatic moments, but it was not simply a matter of lurches on days such as 'Black Thursday' or 'Black Monday'. It was best understood as a long downward grind, punctuated by brief rallies, all the way from 380 points in September 1929 to just over 40 points by the summer of 1932. If Fisher had cut his losses and stepped back from the market in late 1929, he would have been fine. He could have returned to his academic research and his many other enthusiasms, and his luxurious lifestyle funded by many years of trading profits along with his income as an author and businessman.

Instead, Fisher doubled down on his initial views. He was convinced the market would turn upward again. He made several comments about how the crash was the 'shaking out of the lunatic fringe' and reflected 'the psychology of panic'. He publicly declared that recovery was imminent. It was not.

Most important, he didn't just stay invested in the market. His confidence that he was right made him continue to rely on borrowed money in the hope of bigger gains. One of Fisher's major investments was in Remington Rand, following the sale of his Rolodex system, 'Index Visible'. The share price tells the story: $58 before the crash, $28 within a few months. Fisher might have learned by then that leverage was terribly risky. But no: he borrowed more money to invest – and the share price soon dropped to $1. That is a sure route to ruin.

We shouldn't be too quick to judge Fisher. Even if you're the smartest one in the room – and Irving Fisher usually was – it simply isn't easy to change your mind.

Irving Fisher's contemporary, Robert Millikan, was no less distinguished a man than Fisher. His interests were a little different, however: Millikan was a physicist. In 1923, as Fisher's stock tips were being devoured across the United States, Millikan was collecting a Nobel Prize.

For all his achievements, Millikan is most famous for an experiment so simple that a school kid can attempt it: the 'oil drop' experiment, in which a mist of oil droplets from a perfume spritzer is given an electrical charge while floating between two electrified plates. Millikan could adjust the voltage between the plates until oil drops were suspended, without moving – and since he could measure the diameter of the drops, he could calculate their mass, and thus also the electrical charge that was precisely offsetting the pull of gravity. This, in effect, allowed Millikan to calculate the electrical charge of a single electron.

I was one of countless students who attempted this experiment in school, but in all honesty I was unable to get my results quite as neat as Millikan's. There are a lot of details to get right – in particular, the experiment depends on correctly measuring the diameter of the tiny oil droplet. Mis-measure that, and all your other calculations will be off.

We now know that even Millikan didn't get his answers quite as neat as he claimed he did. He systematically omitted observations that didn't suit him, and lied about those omissions. (He also minimised the contribution of a junior colleague, Harvey Fletcher.) Historians of science argue about the seriousness of this cherry-picking, ethically and practically. What seems clear is that if the scientific world had seen all of Millikan's results, it would have had less confidence that his answer was right. That would have been no bad thing, because it wasn't. Millikan's answer was too low.[6]

The charismatic Nobel laureate Richard Feynman pointed out in the early 1970s that the process of fixing Millikan's error with better measurements was a strange one: 'One is a little bit bigger than Millikan's, and the next one's a little bit bigger than that, and the next one's a little bit bigger than that, until finally they settle down to a number which is higher. Why didn't they discover the new number was higher right away?'[7]

The answer is that whenever a number was close to Millikan's, it was accepted without too much scrutiny. When a number seemed wrong it would be viewed with scepticism. Reasons would be found to discard it. As we saw in the first chapter, our preconceptions are powerful things. We filter new information. If it accords with what we expect, we'll be more likely to accept it.

And since Millikan's estimate was too low, it would be rare to have a measurement that was so much lower as to be unexpected. Typically, the surprising measurements would be substantially larger than Millikan's instead. Accepting them was a long and gradual process. It wasn't helped by the fact that Millikan had discarded some of his measurements to make himself seem like a more accomplished scientist. But we can be confident that it would have happened anyway, because a later study found the same pattern of gradual convergence in other estimates of physical constants such as Avogadro's number and Planck's constant.* Convergence continued throughout the 1950s and 1960s and sometimes into the 1970s.[8] It's a powerful demonstration of the way that even scientists measuring essential and unchanging facts filter the data to suit their preconceptions.

This shouldn't be entirely surprising. Our brains are always trying to make sense of the world around us, based on incomplete information. The brain makes predictions about what it expects, and tends to fill in the gaps, often based on surprisingly sparse data. That is why we can understand a routine telephone conversation on a bad line – until the point at which genuinely novel information such as a phone number or street address is being spoken through the static. Our brains fill in the gaps – which is why we see what we expect to see, and hear what we expect to hear, just as Millikan's successors found what they expected to find. It is only when we can't fill in the gaps that we realise just how bad the connection is.

We even smell what we expect to smell. When scientists give people a whiff of scent, the reactions differ sharply based on whether the scientists have told the experimental subjects 'this is the aroma of a gourmet cheese' or 'this is the stink of armpits'.[9] (It's both: they're smelling an aromatic molecule present in both runny cheese and bodily crevices.)

This process of sensing what you expect to sense is widespread. In the cheese study, it was visceral. In the case of the electron charge or Avogadro's number, it was cerebral. In both cases, it seems to have been unconscious.

But we can also filter new information consciously, because we don't want it to spoil our day. Back in the first chapter,

---

* I'll spare you my efforts to define these physical constants. For our purposes what matters is that they are hard to measure precisely, and that each attempt to improve the accuracy of these measurements seems to have been systematically swayed by previous attempts.

we encountered students who'd pay *not* to have their blood tested for herpes and investors who avoided checking their stock portfolios when the news might be bad. Here's another example – a study published in 1967, which asked undergraduates to listen to tape-recorded speeches and requested that they 'judge the persuasiveness and sincerity of talks prepared by high school juniors and seniors . . . After each talk you will be given a rating sheet to rate the persuasiveness and sincerity of the speech.'

However, there was a catch. The talks were clouded with annoying static. The experimental subjects were told: 'Since the talks were recorded on a small portable tape recorder there is considerable electrical interference. The interference can be "adjusted out" by pressing and then immediately releasing the control button. Use of the control several times in a row reduces somewhat the static and other interference noise.'[10]

Fine. Of course, as you can guess by now, the experiment involved some deception. Some of the undergraduates were committed Christians, and others were committed smokers. One of the talks was based on an old-school atheistic pamphlet titled *Christianity Is Evil*, another relied on 'an authoritative refutation of the arguments linking smoking to lung cancer', and a third spoke with similar authority about the fact that smoking *did* cause lung cancer.

As we've seen, all of us are capable of metaphorically filtering the information that comes our way, discarding some ideas and clinging on to others. In this experiment, the filter was more literal: static that obscured the messages the experimental subjects were supposed to listen to and evaluate. Pressing a button could remove the crackle and hiss – but not everyone enthusiastically mashed the button for every speech. It may not surprise you to hear that the Christians were content to leave the militant atheism behind a reassuring fog of static. Smokers pressed the button repeatedly to listen to the talk explaining that their habit was perfectly safe, while allowing the static to float back in when a different taped message told them unwelcome news.

One of the reasons facts don't always change our minds is that we are keen to avoid uncomfortable truths. These days, of course, we don't need to mess around with a static-reducing button. On social media we can choose who to follow and who to block. A vast range of cable channels, podcasts and streaming video lets us decide what to watch and what to ignore. We have more such choices than ever before, and you can bet that we'll use them.

If you do have to absorb unwelcome facts, not to worry: you can always selectively misremember them. That was the conclusion of Baruch Fischhoff and Ruth Beyth, two psychologists who ran an elegant experiment in 1972. They conducted a survey in which they asked male and female students for predictions about Richard Nixon's imminent presidential visit to China and the Soviet Union. How likely was it that Nixon and Mao Zedong would meet? What were the chances that the US would grant diplomatic recognition to China? Would the US and USSR announce a joint space programme?

Fischhoff and Beyth wanted to know how people would later remember their forecasts. They'd given their subjects every chance, since the forecasts were both specific and written down. (Usually our forecasts are rather vague prognostications in the middle of conversation. We rarely commit them to writing.) So one might have hoped for accuracy. But no – the subjects flattered themselves hopelessly. If they put some event at a 25 per cent likelihood, and then it happened, they might then remember they'd called it as a 50/50 proposition. If a subject had put a 60 per cent probability on an event which later failed to happen, she might later recall that she'd forecast a 30 per cent probability. The Fischhoff-Beyth paper was titled 'I knew it would happen'.

It's yet another striking illustration of how our emotions lead us to filter the most straightforward information – our own memory of an estimate we made not long ago, and went to the trouble of committing to paper.[11] In some ways, this shows a remarkable mental flexibility. But rather than admit error and learn from it, Fischhoff and Beyth's subjects were changing their own recollections to ensure that no painful reckoning with reality was required. As we've seen: admitting you're wrong, then changing your view, is not an easy thing to do.

Of course, Irving Fisher wouldn't have had to change his mind if he'd been right all along. Perhaps his real downfall was not the failure to adjust, but the failure to forecast accurately in the first place? Perhaps. It is certainly preferable to be right first time than to learn through painful experience. But the best studies we have of forecasting ability suggest that being right first time isn't easy either.

In 1987, a young Canadian-born psychologist, Philip Tetlock, planted a time bomb under the forecasting industry that would not explode for eighteen years. Tetlock had been part of a rather grand project in which social scientists had been tasked with preventing nuclear war between the US and the USSR. As part of that project, he had interviewed many top experts to get their sense of what was happening in the Soviet Union, how the Soviets might respond to Ronald Reagan's hawkish stance, what might happen next, and why.

But he found himself frustrated: frustrated by the fact that the leading political scientists, Sovietologists, historians and policy wonks had such contradictory views about what might happen next; frustrated by their refusal to change their minds in the face of contradictory evidence; and frustrated by the many ways in which even failed forecasts could be justified. Some predicted disaster, but were happy to rationalise the lack of catastrophe: 'I was nearly right but fortunately it was Gorbachev rather than some neo-Stalinist who took over the reins.' 'I made the right mistake: far more dangerous to underestimate the Soviet threat than overestimate it.' Or, of course, the get-out for all failed stock market forecasts, 'Only my timing was wrong.'

Tetlock's response was patient, painstaking and quietly brilliant. Following in the footsteps of Fischhoff and Beyth, but with more detail and on a much larger scale, he began to collect forecasts from almost three hundred experts, eventually accumulating 27,500 predictions. The main focus of the questions he asked was on politics and geopolitics, with a few from other areas such as economics thrown in. Tetlock sought clearly defined questions, enabling him with the benefit of hindsight to pronounce each forecast right or wrong. Then he simply waited while the results rolled in – for eighteen years.

Tetlock published his conclusions in 2005, in a subtle and scholarly book, *Expert Political Judgment*. He found that his experts were terrible forecasters. This was true in both the simple sense that the forecasts failed to materialise and in the deeper sense that the experts had little idea of how confident they should be in making forecasts in different contexts. It is easier to make forecasts about the territorial integrity of Canada than about the territorial integrity of Syria but, beyond the most obvious cases, the experts Tetlock consulted failed to distinguish the Canadas from the Syrias. Tetlock's experts, like Fischhoff and Beyth's amateurs, also dramatically misremembered their own forecasts, recalling some of their failures as things they'd been right about all along.[12]

Adding to the appeal of this tale of expert hubris, Tetlock found that the most famous experts made even less accurate forecasts than those outside the media spotlight. Other than that, the humiliation was evenly distributed. Regardless of political ideology, profession and academic training, experts failed to see into the future.

Most people, hearing about Tetlock's research, simply conclude that either the world is too complex to forecast, or that experts are too stupid to forecast it, or both. But there was one person who kept faith in the possibility that even for intractable human questions of macroeconomics and geopolitics, a forecasting approach might exist that would bear fruit. That person was Philip Tetlock himself.

In 2013, on the auspicious date of 1 April, I received an email from Tetlock inviting me to join what he described as 'a major new research programme funded in part by Intelligence Advanced Research Projects Activity, an agency within the US intelligence community'.

The core of the programme, which had been running since 2011, was a collection of quantifiable forecasts much like Tetlock's long-running study. The forecasts would be of economic and geopolitical events, 'real and pressing matters of the sort that concern the intelligence community – whether Greece will default, whether there will be a military strike on Iran, etc'. These forecasts took the form of a tournament with thousands of contestants; the tournament ran for four annual seasons.

'You would simply log on to a website,' Tetlock's email continued, 'give your best judgment about matters you may be following anyway, and update that judgment if and when you feel it should be. When time passes and forecasts are judged, you could compare your results with those of others.'

I did not participate. I told myself I was too busy; perhaps I was too much of a coward as well. But the truth is that I did not participate because, largely thanks to Tetlock's work, I had concluded that the forecasting task was impossible.

Still, more than 20,000 people embraced the idea. Some could reasonably be described as having some professional standing, with experience in intelligence analysis, think-tanks or academia. Others were pure amateurs. Tetlock and two other psychologists, Barbara Mellers (Mellers and Tetlock are married) and Don Moore, ran experiments with the co-operation of this army of volunteers. Some were given training in some basic statistical techniques (more on this in a moment); some were assembled into teams; some were given information about other forecasts; while others operated in isolation. The entire exercise was given the name of the Good Judgment Project, and the aim was to find better ways to see into the future.

This vast project has produced a number of insights, but the most striking is that there was a select group of people whose forecasts, while by no means perfect, were vastly better than the dart-throwing-chimp standard reached by the typical prognosticator. What is more, they got better over time rather than fading away as their luck changed. Tetlock, with an uncharacteristic touch of hyperbole, called them 'superforecasters'.

The cynics were too hasty: it is possible to see into the future after all.

What makes a superforecaster? Not subject-matter expertise: professors were no better than well-informed amateurs. Nor was it a matter of intelligence, otherwise Irving Fisher would have been just fine. But there were a few common traits among the better forecasters.

First, encouragingly for us nerds, it did help to have some training – of a particular kind. Just an hour of training in basic statistics improved the performance of forecasters by helping them turn their expertise about the world into a sensible probabilistic forecast, such as 'the chance that a woman will be elected President of the US within the next ten years is 25 per cent'. The tip that seemed to help most was to encourage them to focus on something called 'base rates'.[13]

What on earth are base rates? Well, imagine that you find yourself at a wedding, sitting at one of the back tables with the drunk schoolfriends of the groom or the disgruntled ex-boyfriend of the bride. (Yes, that sort of wedding.) At a tedious moment during one of the speeches, the conversation at your table turns to the distasteful question: will these two actually make it? Will the marriage last or is the relationship doomed to divorce?

The instinctive starting point is to think about the couple. It's always hard to imagine divorce in the middle of the romance of a wedding day (although sharing a whisky with the bride's ex-boyfriend may shake you out of that rosy glow) but you'd naturally ponder questions such as: 'Do they seem happy and committed to each other?'; 'Have I ever seen them argue?'; and 'Have they split up and got back together three times already?' In other words, we make a forecast with the facts that are in front of our nose.

But it is a better idea to zoom out and find one very straightforward statistic*: in general, how many marriages

---

* It's naughty of me to call this a 'straightforward' statistic. In the UK, according to the Office for National Statistics (Statistical Release, 29 November 2019), 22 per cent of 1965 marriages had ended in divorce by 1985. That figure has risen over time: 38 per cent of 1995 marriages had ended in divorce by 2015. There is now evidence that the divorce rate is falling again – but it is obviously too early to say how many recent marriages will last twenty years. Evidently it's a matter of judgement – and the available data – as to which base rate you think is relevant. All UK marriages? All recent marriages? All marriages between people of a certain age, or education level? It's not straightforward at all, if I am honest. But it is better to try to find a relevant base rate and reason from there, than to pull numbers out of your brain without any context.

end in divorce? This number is known as the 'base rate'. Unless you know whether the base rate is 5 per cent or 50 per cent, all the gossip you're getting from the grumpy ex doesn't fit into any useful framework.

The importance of the base rate was made famous by the psychologist Daniel Kahneman, who coined the phrase 'the outside view and the inside view'. The inside view means looking at the specific case in front of you: this couple. The outside view requires you to look at a more general 'comparison class' of cases – here, the comparison class is all married couples. (The outside view needn't be statistical, but it often will be.)

Ideally, a decision-maker or a forecaster will combine the outside view and the inside view – or, similarly, statistics plus personal experience. But it's much better to start with the statistical view, the outside view, and then modify it in the light of personal experience than it is to go the other way around. If you start with the inside view you have no real frame of reference, no sense of scale – and can easily come up with a probability that is ten times too large, or ten times too small.

Second, keeping score was important. As Tetlock's intellectual predecessors Fischhoff and Beyth had demonstrated, we find it challenging to do something as simple as remembering whether our earlier forecasts were right or wrong.

Third, superforecasters tended to update their forecasts frequently as new information emerged, which suggests that a receptiveness to new evidence was important. This willingness to adjust predictions is correlated with making better predictions in the first place: it wasn't just that the superforecasters beat the others because they were news junkies with too much time on their hands, prospering by endlessly tweaking their forecasts with each new headline. Even if the tournament rules had demanded a one-shot forecast, the superforecasters would have come top of the heap.

Which points to the fourth and perhaps most crucial element: superforecasting is a matter of having an open-minded personality. The superforecasters are what psychologists call 'actively open-minded thinkers' – people who don't cling too tightly to a single approach, are comfortable abandoning an old view in the light of fresh evidence or new arguments, and embrace disagreements with others as an opportunity to learn. 'For superforecasters, beliefs are hypotheses to be tested, not treasures to be guarded,' wrote Philip Tetlock after the study had been completed. 'It would be facile to reduce superforecasting to a bumper-sticker slogan, but if I had to, that would be it.'[14]

And if even that is too long for the bumper sticker, what about this: superforecasting means being willing to change your mind.

The unfortunate Irving Fisher had struggled to change his mind. Not everyone had the same difficulty. The contrast with John Maynard Keynes is striking, despite the many similarities the two men shared. Keynes, like Fisher, was a colossal figure in economics. Like Fisher, he was a popular author, a regular newspaper commentator, a friend of powerful politicians, and a charismatic speaker. (After witnessing Keynes giving a speech, the Canadian diplomat Douglas LePan was moved to write, 'I am spellbound. This is the most beautiful creature I have ever listened to. Does he belong to our species? Or is he from some other order?')[15] And like Fisher, Keynes was an enthusiastic participant in financial markets – founding an early hedge fund, dabbling in currency speculation, and managing a large portfolio on behalf of King's College, Cambridge. His ultimate fate, however, was very different. The similarities and the contrasts between the two men are instructive.

Unlike Fisher, who had had to scramble for his success, Keynes was the ultimate insider. As a schoolboy Keynes was educated at Eton College – just like Britain's first Prime Minister, and nineteen others since. Like his father, he became a senior academic: a Fellow of King's College, the most spectacular of all the Cambridge colleges. His job during the First World War was managing both debt and currency on behalf of the British Empire; he'd barely turned thirty. He knew everyone. He whispered in the ear of Prime Ministers. He had the inside track on whatever was going on in the British economy – the Bank of England would even call him to give him advance notice of interest rate movements.

But this child of the British establishment was a very different person to his American counterpart. He loved fine wines and rich food; he gambled at Monte Carlo. His sex life was more like that of a 1970s pop star than a 1900s economist: bisexual, polyamorous, eventually settling down not with his childhood sweetheart but with a Russian ballerina, Lydia Lopokova. One of Keynes's ex-boyfriends was the best man at their wedding.

He was adventurous in other ways, too. In 1918, for example, Keynes worked at the British Treasury. The First World War was still raging. The German army was camped outside Paris, shelling the city. But Keynes caught wind of the fact that, in Paris, the great French impressionist artist Edgar Degas was about to auction his vast collection of pieces by France's greatest nineteenth-century painters: Manet, Ingres and Delacroix.[16]

And so Keynes launched an insane adventure. First, he persuaded the British Treasury, which was four years into fighting the most devastating war the planet had yet seen, to put together a fund for purchasing art of £20,000 – millions in today's money. There was certainly a logic to the idea that it was a buyer's market, but you've got to be pretty persuasive to free up funds from a wartime treasury to splurge on nineteenth-century French art.

Then, escorted by destroyers and a silver airship, Keynes crossed the Channel to France with the director of London's National Gallery, who was wearing a fake moustache so that nobody recognised him. With the German artillery booming beyond the horizon, they showed up at the auction and cleared Degas out. The National Gallery got twenty-seven masterpieces at rock-bottom prices. Keynes even bought a few for himself.

After escaping back across the Channel, and exhausted after his adventures in Paris, Keynes showed up at the door of his friend Vanessa Bell and told her that he'd left a Cézanne outside in the hedge – could he please have a hand carrying it in? (Bell was the sister of the author Virginia Woolf and the lover of Keynes's ex-boyfriend Duncan Grant, although she was married to someone else ... Keynes's social circle was complicated.) Keynes had got himself a bargain: these days a good Cézanne is worth a lot more than anything the National Gallery dared to purchase at the auction. But what Irving Fisher would have made of it all, I do not know.

At the end of the war, Keynes represented the British Treasury at the peace conference in Versailles. (He was disgusted at the outcome – and subsequent events proved him right.) Then, with currencies free-floating and volatile, Keynes set up what some historians describe as the first hedge fund to speculate on their movements. He raised capital from rich friends, and from his own father, to whom he made the not entirely reassuring comment, 'Win or lose, this high-stakes gambling amuses me!'

Initially Keynes made money fast – over £25,000, even more than the art fund he'd wheedled out of the Treasury. His bet, in brief, was that the currencies of France, Italy and Germany would suffer in a bout of post-war inflation. In this he was broadly correct. Yet there's an old saying, often attributed (without evidence) to Keynes himself: the market can stay wrong longer than you can stay solvent. A brief surge of optimism about Germany's prospects wiped out Keynes's fund in 1920. Undaunted, he went back to his investors. 'I am not in a position to risk any capital myself, having quite exhausted my resources,' he noted. But the spellbinding Keynes persuaded others to invest and his fund was back in profit by 1922.

One of Keynes's next investment projects – he had several – concerned the portfolio of King's College, Cambridge. Five centuries old, the college had long-standing rules on its investment policy, leaving it reliant on agricultural rents and very conservative investments such as railway bonds and government gilts. In 1921 the ever-persuasive Keynes convinced the college to change these rules to give him complete discretion over a significant slice of the college portfolio.

Keynes's strategy for this money was top-down. He would forecast booms and recessions both in the UK and abroad, and invest in shares and commodities accordingly, moving across different sectors and countries depending on the macroeconomic outlook.

Such an approach seemed to make sense. Keynes was the leading economic theorist in the country. He was receiving tips from the Bank of England. If anyone could call the ebb and flow of the British economy, it was John Maynard Keynes.

If.

Keynes, like Fisher, did not predict the great crash of 1929. Unlike Fisher, though, he recovered. Keynes died a millionaire, his reputation enhanced by his financial acumen. The reason is simple: Keynes, unlike Fisher, changed his mind, and his investment strategy.

Keynes had one advantage over Fisher: his track record as an investor had been painfully mixed. Yes, he had scored a remarkable coup in the art auction of 1918, and made a small fortune in the currency markets in 1922. But he had been wiped out in 1920, and his clever-seeming approach with the King's College portfolio wasn't working either. Over the course of the 1920s, Keynes's attempts to forecast the business cycle had led him to trail the market as a whole by about 20 per cent. That is not a disaster, but it is certainly an indication that all is not well.

None of this helped Keynes see the great crash of 1929, but it did help him react to it. He had already been pondering his limitations as an investor, and wondering whether a different approach might pay off. When the crash hit, Keynes shrugged, and adjusted.

By the early 1930s, Keynes had abandoned business-cycle forecasting entirely. The greatest economist in the world had decided that he just couldn't do it well enough to make money. It is a striking instance of humility from a man famous for his self-confidence. But Keynes had looked at the evidence and done something unusual: he'd changed his mind.

He moved instead to an investment strategy that required no great macroeconomic insight. Instead, he explained, 'As time goes on, I get more and more convinced that the right method in investment is to put fairly large sums into enterprises which one thinks one knows something about and in the management of which one thoroughly believes.' Forget what the economy is doing; just find well-managed companies, buy some shares, and don't try to be too clever. And if that approach sounds familiar, it's most famously associated with Warren Buffett, the world's richest investor – and a man who loves to quote John Maynard Keynes.

Keynes is rightly viewed today as a successful investor. At King's College, he recovered from the poor performance of the early years. When two financial economists, David Chambers and Elroy Dimson, recently studied Keynes's track record with the King's College portfolio, they found it to be excellent. Keynes secured high returns with modest risks, and outperformed the stock market as a whole by an average of six percentage points a year over a quarter of a century. That's an impressive reward for being able to change your mind.[17]

It all sounds so simple: things are going badly, so do something different. Why, then, did Irving Fisher struggle to adapt?

Fisher's first problem, ironically, was his successful track record. He was seriously wealthy by the end of the 1920s, having prospered in almost every endeavour he had attempted. As an investor, he had correctly predicted the productivity boom of the 1920s and correctly judged that the stock market would soar, and his leveraged bets on those judgements had paid off handsomely. Unlike Keynes, Fisher had received very little evidence of his own fallibility. It must have been hard for him to take in the scale of the financial bloodbath. It was all too tempting to write it off as a brief spasm of lunacy, which is what Fisher did.

In contrast, when the market crashed, Keynes was able to see it – and himself – for what it was. He'd been in crashes before, and lost heavily before. He was like a physicist who'd been forewarned that Robert Millikan's research was flawed, so his estimates shouldn't be taken too seriously; or perhaps like an experimental subject sniffing a test-tube after being told 'this might be cheese, or it might be armpits, so think carefully'.

Fisher was vulnerable in a second way. He was constantly writing about his investment ideas, pinning his reputation to the idea that the stock market was on the up and up. There is a lot of vague prophecy in the forecasting business, so such public commitments are admirably honest. They are also dangerous. It wasn't the concreteness of the predictions that was the problem. As we've seen, superforecasters tend to keep a careful record of their predictions. How else can they learn from their mistakes? No: it was the high public profile that made it harder for Fisher to change his mind.

One study of this, conducted by psychologists Morton Deutsch and Harold Gerard in 1955, asked college students to estimate the lengths of lines – a modification of the experiments conducted by Solomon Asch a few months previously, described in the sixth chapter. Some of the students did not write their estimates down. Others wrote their estimates down on an erasable pad, before erasing the result. Still others wrote their estimates down in permanent marker. As new information emerged, the students who had made this more public commitment were the least willing to change their minds.[18]

'Kurt Lewin noticed [this effect] in the 1930s,' says Philip Tetlock, referring to one of the founders of modern psychology. 'Making public commitments "freezes" attitudes in place. So saying something dumb makes you a bit dumber. It becomes harder to correct yourself.'[19]

And Fisher's commitment could hardly have been more public. Two weeks before the Wall Street crash began, he was reported by the *New York Times* as saying, 'Stocks have reached what looks like a permanently high plateau.' How do you back away from that?

Fisher's third problem – perhaps the deepest – was his belief that, in the end, the future was knowable. 'The sagacious businessman is constantly forecasting,' he once wrote. Maybe. But contrast that with John Maynard Keynes's famous view about long-term forecasts: 'About these matters there is no scientific basis on which to form any calculable probability whatever. We simply do not know.'

Fisher, a man who was happy to specify the perfect angle for the out-turn of the foot, admire the rigour of the 'clo' unit of warming, and estimate the productivity gain from Prohibition, believed that with a sufficiently powerful statistical lens any problem would yield to the man of science. The statistical lens is indeed powerful. Still, I hope that I have convinced you that for any problem, it takes more than mere numbers to make the world add up.

Poor Irving Fisher had believed himself to be a man of logic and reason. He was a campaigner for education reform and the proven benefits of a vegetarian diet, and a student of 'the science of wealth'. And yet he became the most famous financial basket-case in the country.

He kept thinking, and working, producing an incisive account of why the Depression had been so severe – including a painful reckoning with the effect of debt on the economy. But while his economic ideas are still respected today, he became a marginalised figure. He was deep in debt to the taxman and to his brokers, and towards the end of his life, a widower living alone in modest circumstances, he became an easy target for scam artists: he was always looking for the big financial break that would restore his fortune. The mansion was long gone. He avoided bankruptcy, and perhaps even prison, because his late wife's sister covered his debts to the value of tens of millions of dollars in today's terms. It was a kindness, but for the proud Professor Fisher it must have been the ultimate humiliation.

The economic historian Sylvia Nasar wrote of Fisher that

'His optimism, overconfidence and stubbornness betrayed him.'[20] Keynes had plenty of confidence too, but he had also learned the hard way that there are certain facts about the world that do not easily yield to logic. Recall his comment to his father – 'this high-stakes gambling amuses me'. The Monte Carlo gambler knew, all along, that while investing was a fascinating game, it was a game nonetheless, and one should not take an unlucky throw of the dice too much to heart. When his early investment ideas failed, he tried something else. Keynes was able to change his mind; Fisher, alas, could not.

Fisher and Keynes died within a few months of each other, not long after the end of the Second World War. Fisher was a much-diminished figure; Keynes was the most influential economist on the planet, fresh from shaping the World Bank, the IMF and the entire global financial system at the Bretton Woods conference in 1944.

Late in his life, Keynes reflected, 'My only regret is that I have not drunk more champagne in my life.' But he is remembered far more for words that he probably never said. Nevertheless, he lived by them: 'When my information changes, I alter my conclusions. What do you do, sir?'

If only he had taught that lesson to Irving Fisher.

Fisher and Keynes were equally expert, and they had the same statistical information at their fingertips – data they themselves had done much to collect. Just as with Abraham Bredius, the art scholar so cruelly tricked by the forger Han van Meegeren, their fates were determined not by their expertise but by their emotions.

This book has argued that it is possible to gather and to analyse numbers in ways that help us understand the world. But it has also argued that very often we make mistakes not because the data aren't available, but because we refuse to accept what they are telling us. For Irving Fisher, and for many others, the refusal to accept the data was rooted in a refusal to acknowledge that the world had changed.

One of Fisher's rivals, an entrepreneurial forecaster named Roger Babson, explained (not without sympathy) that while Fisher was 'one of the greatest economists in the world today and a most useful and unselfish citizen', he had failed as a forecaster because 'he thinks the world is ruled by figures instead of feelings'.[21]

I hope that this book has persuaded you that it is ruled by both.

THE GOLDEN RULE

Be curious

> I can think of nothing an audience won't understand. The only problem is to interest them; once they are interested they understand anything in the world.
>
> —ORSON WELLES[1]

I've laid down ten statistical commandments in this book.

First, we should learn to stop and notice our emotional reaction to a claim, rather than accepting or rejecting it because of how it makes us feel.

Second, we should look for ways to combine the 'bird's eye' statistical perspective with the 'worm's eye' view from personal experience.

Third, we should look at the labels on the data we're being given, and ask if we understand what's really being described.

Fourth, we should look for comparisons and context, putting any claim into perspective.

Fifth, we should look behind the statistics at where they came from – and what other data might have vanished into obscurity.

Sixth, we should ask who is missing from the data we're being shown, and whether our conclusions might differ if they were included.

Seventh, we should ask tough questions about algorithms and the big datasets that drive them, recognising that without intelligent openness they cannot be trusted.

Eighth, we should pay more attention to the bedrock of official statistics – and the sometimes heroic statisticians who protect it.

Ninth, we should look under the surface of any beautiful graph or chart.

And tenth, we should keep an open mind, asking how we might be mistaken, and whether the facts have changed.

I realise that having ten commandments is something of a cliché. And in truth, they're not commandments so much as rules of thumb, or habits of mind that I've acquired the hard way as I've gone along. You might find them worth a try yourself, when you come across a statistical claim of particular interest to you. Of course, I don't expect you to run personally through the checklist with every claim you see in the media – who has the time for that? Still, they can be useful in forming a preliminary assessment of your news source. Is the journalist making an effort to define terms, provide context, assess sources? The less these habits of mind are in evidence, the louder alarm bells should ring.

Ten rules of thumb is still a lot for anyone to remember, so perhaps I should try to make things simpler. I realise that these suggestions have a common thread – a golden rule, if you like.

Be curious.

Look deeper and ask questions. It's a lot to ask, but I hope that it is not too much. At the start of this book I begged you not to abandon the idea that we can understand the world by looking at it with the help of statistics, in favour of the cynical distrust so temptingly offered by the likes of Darrell Huff. I believe we can – and should – be able to trust that numbers can give us answers to important questions. My colleagues and I at *More or Less* work hard to earn listeners' trust that we're coming to the same conclusions they would if they investigated the issue themselves. But of course we want listeners to be curious and to question us, too. *Nullius in verba.* We shouldn't trust without also asking questions.

The philosopher Onora O'Neill once declared, 'Well-placed trust grows out of active inquiry rather than blind acceptance.'[2] That seems right. If we want to be able to trust the world around us, we need to show an interest and ask a few basic questions. I hope I've persuaded you that those questions aren't obscure or overly technical; they are what any thoughtful, curious person would be happy to ask. And despite all the confusions of the modern world, it has never been easier to find answers to those questions.

Curiosity, it turns out, can be a remarkably powerful thing.

About a decade ago, a Yale University researcher, Dan Kahan, showed students some footage of a protest outside an unidentified building. Some of the students were told that it was a pro-life demonstration outside an abortion clinic. Others were informed that it was a gay rights demonstration outside an army recruitment office. The students were asked some factual questions. Was it a peaceful protest? Did the protesters try to intimidate people passing by? Did they scream or shout? Did they block the entrance to the building?

The answers people gave depended on the political identities they embraced. Conservative students who believed they were looking at a demonstration against abortion saw no problems with the protest: no abuse, no violence, no obstruction. Students on the left who thought they were looking at a gay rights protest reached the same conclusion: the protesters had conducted themselves with dignity and restraint.

But right-wing students who thought they were looking at a gay rights demonstration reached a very different conclusion; as did left-wing students who believed they were watching an anti-abortion protest. Both these groups concluded that the protesters had been aggressive, intimidating and obstructive.[3]

Kahan was studying a problem we met in the first chapter: the way our political and cultural identity – our desire to belong to a community of like-minded, right-thinking people – can, on certain hot-button issues, lead us to reach the conclusions we wish to reach. Depressingly, not only do we reach politically comfortable conclusions when parsing complex statistical claims on issues such as climate change, we reach politically comfortable conclusions regardless of the evidence of our own eyes.*

And, as we saw earlier, expertise is no guarantee against this kind of motivated reasoning: Republicans and Democrats with high levels of scientific literacy are further apart on climate change than those with little scientific education. The same disheartening pattern holds from nuclear power to gun control to fracking: the more scientifically literate opponents are, the more they disagree. The same is true for numeracy. 'The greater the proficiency, the more acute the polarization,' notes Kahan.[4]

After a long and fruitless search for an antidote to tribalism, Kahan could be forgiven for becoming jaded.[5] Yet a few years ago, to his surprise, Kahan and his colleagues stumbled upon a trait that some people have – and that other people

---

* The study is titled 'They Saw a Protest', echoing a classic psychology paper from 1954, 'They Saw a Game', which found similarly biased perceptions when rival fans watched footage of a bad-tempered game of football.

can be encouraged to develop – which inoculates us against this toxic polarisation. On the most politically polluted, tribal questions, where intelligence and education fail, this trait does not.

And if you're desperately, burningly curious to know what it is – congratulations. You may be inoculated already.

Curiosity breaks the relentless pattern. Specifically, Kahan identified 'scientific curiosity'. That's different from scientific literacy. The two qualities are correlated, of course, but there are curious people who know rather little about science (yet), and highly trained people with little appetite to learn more.

More scientifically curious Republicans aren't further apart from Democrats on these polarised issues. If anything, they're slightly closer together. It's important not to exaggerate the effect. Curious Republicans and Democrats still disagree on issues such as climate change – but the more curious they are, the more they converge on what we might call an evidence-based view of the issues in question. Or to put it another way, the more curious we are, the less our tribalism seems to matter. (There is little correlation between scientific curiosity and political affiliation. Happily, there are plenty of curious people across the political spectrum.)

Although the discovery surprised Kahan, it makes sense. As we've seen, one of our stubborn defences against changing our minds is that we're good at filtering out or dismissing unwelcome information. A curious person, however, enjoys being surprised and hungers for the unexpected. He or she will not be filtering out surprising news, because it's far too intriguing.

The scientifically curious people Kahan's team studied were originally identified with simple questions, buried in a marketing survey so that people weren't conscious that their curiosity was being measured. One question, for example, was 'How often do you read science books?' Scientifically curious people are more interested in watching a documentary about space travel or penguins than a basketball game or a celebrity gossip show. And they didn't just answer survey questions differently, they also made different choices in the psychology lab. In one experiment, participants were shown a range of headlines about climate change and invited to pick the 'most interesting' article to read. There were four headlines. Two suggested climate scepticism and two did not; two were framed as surprising and two were not:

1. 'Scientists Find Still More Evidence that Global Warming Actually Slowed in Last Decade' (sceptical, unsurprising)
2. 'Scientists Report Surprising Evidence: Arctic Ice Melting Even Faster than Expected' (surprising and not sceptical)
3. 'Scientists Report Surprising Evidence: Ice *Increasing* in Antarctic, *Not* Currently Contributing to Sea Level Rise' (sceptical and surprising)
4. 'Scientists Find Still More Evidence Linking Global Warming to Extreme Weather' (neither surprising nor sceptical)

Typically we'd expect people to reach for the article that pandered to their prejudices: the Democrats would tend to favour a headline that took global warming seriously while Republicans would prefer something with a sceptical tone. Scientifically curious people – Republicans or Democrats – were different. They were happy to grab an article which ran counter to their preconceptions, as long as it seemed surprising and fresh. And once you're actually reading the article, there's always a chance that it might teach you something.

A surprising statistical claim is a challenge to our existing world-view. It may provoke an emotional response – even a fearful one. Neuroscientific studies suggest that the brain responds in much the same anxious way to facts which threaten our preconceptions as it does to wild animals which threaten our lives.[6] Yet for someone in a curious frame of mind, in contrast, a surprising claim need not provoke anxiety. It can be an engaging mystery, or a puzzle to solve.

A curious person might, at this point, have some questions. When I met Dan Kahan, the question that was most urgent in my mind was – can we cultivate curiosity? Can we become more curious, and can we inspire curiosity in others?

There are reasons to believe that the answers are 'yes'. One reason, says Kahan, is that his measure of curiosity suggests that incremental change is possible. When he measures scientific curiosity, he doesn't find a lump of stubbornly incurious people at one end of the spectrum and a lump of voraciously curious people at the other, with a yawning gap in the middle. Instead, curiosity follows a continuous bell curve: most people are either moderately incurious or moderately curious. This doesn't prove that curiosity can be cultivated; perhaps that bell curve is cast in iron. Yet it does at least hold out some hope that people can be nudged a little further towards the curious end of that curve, because no radical leap is required.

A second reason is that curiosity is often situational. In the right place, at the right time, curiosity will smoulder in any of us.* Indeed, Kahan's discovery that an individual's

---

* Trolls, populists, manufacturers of outrage and other professional controversialists will, of course, try to frame debates in ways that crush curiosity and reinforce preconceptions. But curious and open-minded folk can also frame debates, and we would be wise to take the opposite tack.

scientific curiosity persisted over time was a surprise to some psychologists. They had believed, with some reason, that there was no such thing as a curious person, just a situation that inspired curiosity. In fact it does now seem that people can tend to be curious or incurious. That does not alter the fact that curiosity can be fuelled or dampened by context. We all have it in us to be curious, or not, about different things at different times.

One thing that provokes curiosity is the sense of a gap in our knowledge to be filled. George Loewenstein, a behavioural economist, framed this idea in what has become known as the 'information gap' theory of curiosity. As Loewenstein puts it, curiosity starts to glow when there's a gap 'between what we know and what we want to know'. There's a sweet spot for curiosity: if we know nothing, we ask no questions; if we know everything, we ask no questions either. Curiosity is fuelled once we know enough to know that we do not know.[7]

Alas, all too often we don't even think about what we don't know. There's a beautiful little experiment about our incuriosity, conducted by the psychologists Leonid Rozenblit and Frank Keil. They gave their experimental subjects a simple task: to look through a list of everyday objects such as a flush lavatory, a zip fastener and a bicycle, and to rate their understanding of each object on a scale of one to seven.[8]

After people had written down their ratings, the researchers would gently launch a devastating ambush. They asked the subjects to elaborate. Here's a pen and paper, they would say; please write out your explanation of a flush lavatory in as much detail as you can. By all means include diagrams.

It turns out that this task wasn't as easy as people had thought. People stumbled, struggling to explain the details of everyday mechanisms. They had assumed that those details would readily spring to mind, and they did not. And to their credit, most experimental subjects realised that they'd been lying to themselves. They had felt they understood zip fasteners and lavatories, but when invited to elaborate, they realised they didn't understand at all. When people were asked to reconsider their previous one-to-seven rating, they marked themselves down, acknowledging that their knowledge had been shallower than they'd realised.

Rozenblit and Keil called this 'the illusion of explanatory depth'. The illusion of explanatory depth is a curiosity-killer and a trap. If we think we already understand, why go deeper? Why ask questions? It is striking that it was so easy to get people to pull back from their earlier confidence: all it took was to get them to reflect on the gaps in their knowledge. And as Loewenstein argued, gaps in knowledge fuel curiosity.

There is more at stake here than zip fasteners. Another team of researchers, led by Philip Fernbach and Steven Sloman, authors of *The Knowledge Illusion*, adapted the flush lavatory question to ask about policies such as a cap-and-trade system for carbon emissions, a flat tax, or a proposal to impose unilateral sanctions on Iran. The researchers, importantly, didn't ask people whether or why they were in favour of or against these policies – there's plenty of prior evidence that such questions would lead people to dig in. Instead, Fernbach and his colleagues just asked them the same simple question: please rate your understanding on a scale of one to seven. Then, the same devastating follow-up: please elaborate; tell us exactly what unilateral sanctions are and how a flat tax works. And the same thing happened. People said, yes, they basically understood these policies fairly well. Then when prompted to explain, the illusion was dispelled. They realised that perhaps they didn't really understand at all.[9]

More striking was that when the illusion faded, political polarisation also started to fade. People who would have instinctively described their political opponents as wicked, and who would have gone to the barricades to defend their own ideas, tended to be less strident when forced to admit to themselves that they didn't fully understand what it was they were so passionate about in the first place. The experiment influenced actions as well as words: researchers found that people became less likely to give money to lobby groups or other organisations which supported the positions they had once favoured.[10]

It's a rather beautiful discovery: in a world where so many people seem to hold extreme views with strident certainty, you can deflate somebody's overconfidence and moderate their politics simply by asking them to explain the details. Next time you're in a politically heated argument, try asking your interlocutor not to justify herself, but simply to explain the policy in question. She wants to introduce a universal basic income, or a flat tax, or a points-based immigration system, or 'Medicare for all'. OK: that's interesting. So what exactly does she mean by that? She may learn something as she tries to explain. So may you. And you may both find that you understand a little less, and agree a little more, than you had assumed.

Figuring out the workings of a flush lavatory, or understanding what a cap-and-trade scheme really is, can require some effort. One way to encourage that effort is to embarrass someone by innocently inviting an overconfident answer on a scale of one to seven; but another, kinder, way is to engage their interest. As Orson Welles said, once people are interested they can understand anything in the world.

How to engage people's interest is neither a new problem nor an intractable one. Novelists, screenwriters and comedians have been figuring out this craft for as long as they have existed. They know that we love mysteries, are drawn in by sympathetic characters, enjoy the arc of a good story, and will stick around for anything that makes us laugh. And scientific evidence suggests that Orson Welles was absolutely right: for example, studies in which people were asked to read narratives and non-narrative texts found that they zipped through the narrative at twice the speed, and recalled twice as much information later.[11]

As for humour, consider the case of the comedian Stephen Colbert's 'civics lesson'. Before his current role as the host of *The Late Show*, Colbert presented *The Colbert Report* in character as a blowhard right-wing commentator.* In March 2011, Colbert began a long-running joke in which he explored the role of money in US politics. He decided that he needed to set up a Political Action Committee – a PAC – to raise funds in case he decided to run for President. 'I clearly need a PAC but I have no idea what PACs do,' he explained to a friendly expert on air.

Over the course of the next few weeks, Colbert had PACs – and Super PACs, and 501(c)(4)s – explained to him: from where they could accept donations, up to what limits, with what transparency requirements, and to spend on what. He was to discover that the right combination of fundraising structures could be used to raise almost any amount of money for almost any purpose, with almost no disclosure. 'Clearly (c)(4)s have created an unprecedented, unaccountable, untraceable cash tsunami that will infect every corner of the next election,' he mused. 'And I feel like an idiot for not having one.'

---

* I was once a guest on *The Colbert Report*. Stephen was a gracious host. In the green room, as himself, he explained the basic idea of the show to me: 'I'll be in character, and my character is an idiot.' Then later, after getting into character, 'I'm going to tear you apart, Harford!'

Colbert later learned how to dissolve his fundraising structures and keep the money – without notifying the taxman. By repeatedly returning to the topic and – in character – demanding advice as to how to abuse the electoral rules, Colbert explored campaign finance in far more depth than any news report could have dreamed of doing.

Did all of this actually improve viewers' knowledge of the issue? It seems so. A team including Kathleen Hall Jamieson, who also worked with Dan Kahan on the scientific curiosity research, used the Colbert storyline to investigate how much people learned amid the laughter. They found that watching *The Colbert Report* was correlated with increased knowledge about Super PACs and 501(c)(4) groups – how they worked, what they could legally do. Reading a newspaper or listening to talk radio also helped, but the effect of *The Colbert Report* was much bigger. One day a week of watching Colbert taught people as much about campaign finance as four days a week reading a newspaper, for example – or five extra years of schooling.

Of course this is a measure of correlation, not causation. It's possible that the people who were already interested in Super PACs tuned in to Colbert to hear him wisecrack about them. Or perhaps politics junkies know about Super PACs and also love watching Colbert. But I suspect the show did cause the growing understanding, because Colbert really did go deep into the details. And large audiences stuck with him – because he was funny.[12]

You don't have to be one of America's best-loved comedians to pull off this trick. The NPR podcast *Planet Money* once shed light on the details of the global economy by designing, manufacturing and importing several thousand T-shirts. This allowed a long-running storyline investigating cotton farming; the role of automation in textiles; how African communities make new fashions out of donated American T-shirts; the logistics of the shipping industry; and strange details such as the fact that the men's shirts, which were made in Bangladesh, attract a tariff of 16.5 per cent, whereas the women's shirts, made in Colombia, are duty-free.[13]

These examples should be models for communication, precisely because they inspire curiosity. 'How does money influence politics?' is not an especially engaging question, but 'If I were running for President, how would I raise lots of money with few conditions and no scrutiny?' is much more intriguing.

Those of us in the business of communicating ideas need to go beyond the fact-check and the statistical smackdown. Facts are valuable things, and so is fact-checking. But if we really want people to understand complex issues, we need to engage their curiosity. If people are curious, they will learn.*

I've found this in my own work with the team who make *More or Less* for the BBC. The programme is often regarded affectionately as a myth-buster, but I feel that our best work is when we use statistics to illuminate the truth rather than to debunk a stream of falsehoods. We try to bring people along with us as we explore the world around us with the help of reliable numbers. What's false is interesting – but not as interesting as what's true.

After the referendum of 2016, in which my fellow British voters decided to leave the European Union, the economics profession engaged in some soul-searching. Most technical experts thought that leaving the EU was a bad idea – costly,

* And if people are not curious, they will not learn until their curiosity can somehow be sparked into life. Remember the TV producers who were making a programme about why inequality had risen, but apparently weren't curious enough to check whether it had?

complex, and unlikely to deliver many of the promised benefits or to solve the country's most pressing problems. Yet, as one infamous soundbite put it, 'the people in this country have had enough of experts'.* Few people seemed to care what economists had to say on the subject, and – to our credit, I think – professional economists wanted to understand what we had done wrong and whether we might do better in future.

Later, at a conference about 'the profession and the public', the great and the good of the British economics community pondered the problem and discussed solutions.[14] We needed to be more chatty and approachable on Twitter, suggested one analysis. We needed to express ourselves clearly and without jargon, offered many speakers – not unreasonably.

My own perspective was slightly different. I argued that we were operating in a politically polarised environment, in which almost any opinion we might offer would be fiercely contested by partisans. Economists deal with controversial issues such as inequality, taxation, public spending, climate change, trade, immigration and, of course, Brexit. In such a febrile environment, speaking slowly and clearly will only get you so far. To communicate complex ideas, we needed to spark people's curiosity – even inspire a sense of wonder. The great science communicators, after all – people such as Stephen Hawking and David Attenborough – do not win over people simply by using small words, crisply spoken. They stoke the flames of our curiosity, making us burn with desire to learn more. If we economists want people to understand economics, we must first engage their interest.

What is true of economists is equally true for scientists,

* When pro-Brexit campaigner Michael Gove said this, he was referring specifically to experts based at international organisations such as the International Monetary Fund. The statement, however, took on a life of its own.

social scientists, historians, statisticians or anyone else with complex ideas to convey. Whether the topic is the evolution of black holes or the emergence of Black Lives Matter, the possibility of precognition or the necessity of preregistration, the details matter – and presented in the right way, they should always have the capacity to fascinate us.

Awaken our sense of wonder, I say to my fellow nerd-communicators. Ignite the spark of curiosity and give it some fuel, using the time-honoured methods of storytelling, character, suspense and humour. But let's not rely on the journalists and the scientists and the other communicators of complex ideas. We have to be responsible for our own sense of curiosity. As the saying goes, 'only boring people get bored'. The world is so much more interesting if we take an active interest in it.

'The cure for boredom is curiosity,' goes an old saying. 'There is no cure for curiosity.'[15] Just so: once we start to peer beneath the surface of things, become aware of the gaps in our knowledge, and treat each question as the path to a better question, we find that curiosity is habit-forming.

Sometimes we need to think like Darrell Huff; there is a place in life for the mean-minded, hard-nosed scepticism that asks, Where's the trick? Why is this lying bastard lying to me?[16] But while 'I don't believe it' is sometimes the right starting point when confronted with a surprising statistical claim, it is a lazy and depressing place to finish.

And I hope you won't finish there. I hope that I have persuaded you that we should make more room both for the novelty-seeking curiosity that says 'tell me more', and the dogged curiosity that drove Austin Bradford Hill and Richard Doll to ask why so many people were dying of lung cancer, and whether cigarettes might be to blame.

If we want to make the world add up, we need to ask questions – open-minded, genuine questions. And once we start asking them, we may find it is delightfully difficult to stop.

Notes

Introduction: How to lie with statistics

1. Umberto Eco, *Serendipities: Language and Lunacy*, London: Hachette, 2015.
2. Robert Matthews, 'Storks Deliver Babies (p = 0.008)', *Teaching Statistics*, 22(2), June 2000, 36–8, http://dx.doi.org/10.1111/1467-9639.00013. Research papers in social science typically say that a relationship is 'statistically significant' if p = 0.05, which means that *if* there was no relationship at all, a pattern at least as clear as the one observed would occur just one in twenty times. The stork paper boasted p = 0.008, which means that *if* there was in fact no relationship between storks and births, a pattern as clear as the one observed would occur just one in 125 times. The tradition of applying such a statistical significance test is regrettable, for reasons we shall not go into now.
3. Conrad Keating, *Smoking Kills*, Oxford: Signal Books, 2009, p.xv.
4. Science Museum, Sir Austin Bradford Hill, http://broughttolife.sciencemuseum.org.uk/broughttolife/people/austinhill; Peter Armitage, 'Obituary: Sir Austin Bradford Hill, 1897–1991', *Journal of the Royal Statistical Society*, Series A (Statistics in Society), 154(3), 1991, 482–4, www.jstor.org/stable/2983156
5. Keating, *Smoking Kills*, pp.85–90.
6. Ibid., p.113.
7. John P.A. Ioannidis, 'A fiasco in the making?' *Stat,* 17 March 2020, https://www.statnews.com/2020/03/17/a-fiasco-in-the-making-as-the-coronavirus-pandemic-takes-hold-we-are-making-decisions-without-reliable-data/
8. 'Taiwan says WHO failed to act on coronavirus transmission warning',*Financial Times,* 20 March 2020, https://www.ft.com/content/2a70a02a-644a-11ea-a6cd-df28cc3c6a68
9. Demetri Sevastopulo and Hannah Kuchler, 'Donald Trump's chaotic coronavirus crisis', *Financial Times,* 27 March 2020, https://www.ft.com/content/80aa0b58-7010-11ea-9bca-bf503995cd6f
10. David Card, 'Origins of the Unemployment Rate: The Lasting Legacy of Measurement without Theory', UC Berkeley and NBER Working

Paper, February 2011, http://davidcard.berkeley.edu/papers/origins-of-unemployment.pdf

11 Naomi Oreskes and Eric Conway, *Merchants of Doubt*, London: Bloomsbury, 2010, Chapter 1; and Robert Proctor, *Golden Holocaust*, Berkeley and Los Angeles: University of California Press, 2011.

12 Smoking And Health Proposal, Brown and Williamson internal memo, 1969 https://www.industrydocuments.ucsf.edu/tobacco/docs/#id=psdw0147.

13 Kari Edwards and Edward Smith, 'A Disconfirmation Bias in the Evaluation of Arguments', *Journal of Personality and Social Psychology*, 71(1), 1996, 5–24.

14 Oreskes and Conway, *Merchants of Doubt*.

15 Michael Lewis, 'Has Anyone Seen the President?', Bloomberg, 9 February 2018, https://www.bloomberg.com/opinion/articles/2018-02-09/has-anyone-seen-the-president

16 Brendan Nyhan, 'Why Fears of Fake News Are Overhyped', *Medium*, 4 February 2019; and Gillian Tett, 'The Kids Are Alright: The Truth About Fake News', *Financial Times*, 6 February 2019, https://www.ft.com/content/d8f43574-29a1-11e9-a5ab-ff8ef2b976c7?desktop=true&segmentId=7c8f09b9-9b61-4fbb-9430-9208a9e233c8

17 CQ Quarterly: https://library.cqpress.com/cqalmanac/document.php?id=cqal65-1259268; and Alex Reinhart, 'Huff and Puff', *Significance*, 11 (4), 2014.

18 Andrew Gelman, 'Statistics for Cigarette Sellers', *Chance*, 25(3), 2012; Reinhart, 'Huff and Puff'.

19 *How to Lie with Smoking Statistics* is stored in the Tobacco Industry Documents library. Alex Reinhart pieced together the manuscript and various documents pertaining to the project: Reinhart, 'The History of "How To Lie With Smoking Statistics"', https://www.refsmmat.com/articles/smoking-statistics.html

20 Suzana Herculano-Houzel, 'What is so special about the human brain?', talk at TED.com given in 2013: https://www.ted.com/talks/suzana_herculano_houzel_what_is_so_special_about_the_human_brain/transcript?ga_source=embed&ga_medium=embed&ga_campaign=embedT

21 On Galileo's telescope: https://thonyc.wordpress.com/2012/08/23/refusing-to-look/; and https://www.wired.com/2008/10/how-the-telesco/; and https://thekindlyones.org/2010/10/13/refusing-to-look-through-galileos-telescope/

Rule One: Search your feelings

1 Also known as *Star Wars: Episode V*; screenplay by Leigh Brackett and Lawrence Kasdan.

2 The van Meegeren case is described in John Godley, *The Master Forger*, London: Home and Van Thal, 1951; and *Van Meegeren: A Case History*, London: Nelson, 1967; Noah Charney, *The Art of Forgery: The Minds, Motives and Methods of Master Forgers, London:* Phaidon, 2015; Frank Wynne, *I Was Vermeer*, London: Bloomsbury, 2007; the BBC TV programme

Fake or Fortune (Series 1, Programme 3, 2011); a series of blog posts by Errol Morris titled 'Bamboozling Ourselves' starting on the *New York Times* website, 20 May 2009; the Boijmans Museum film *Van Meegeren's Fake Vermeers* (2010, available on YouTube at https://www.youtube.com/watch?v=NnnkuOz08GQ); and particularly Jonathan Lopez, *The Man Who Made Vermeers*, London: Houghton Mifflin, 2009.

3 Different accounts exist of exactly how van Meegeren made this confession – another account has van Meegeren equating himself with the Dutch master more directly: 'The painting in Göring's hands is not, as you assume, a Vermeer of Delft, but a van Meegeren!' The quotation in the text is from Frank Wynne's book *I was Vermeer.*

4 Ziva Kunda, 'Motivated Inference: Self-Serving Generation and Evaluation of Causal Theories', *Journal of Personality and Social Psychology*, 53(4), 1987, 636–47.

5 Stephen Jay Gould, 'The median isn't the message', *Discover* 6 June 1985, 40–2.

6 This experiment was described on NPR's 'The Hidden Brain' podcast: *You 2.0: The Ostrich Effect*, 6 August 2018, https://www.npr.org/templates/transcript/transcript.php?storyId=636133086

7 Nachum Sicherman, George Loewenstein, Duane J. Seppi, Stephen P. Utkus, 'Financial Attention', *Review of Financial Studies*, 29(4), 1 April 2016, 863–97, https://doi.org/10.1093/rfs/hhv073

8 'Viral post about someone's uncle's coronavirus advice is not all it's cracked up to be', *Full Fact,* 5 March 2020, https://fullfact.org/online/coronavirus-claims-symptoms-viral/

9 Guy Mayraz, 'Wishful Thinking', 25 October 2011, http://dx.doi.org/10.2139/ssrn.1955644

10 Linda Babcock and George Loewenstein, 'Explaining Bargaining Impasse: The Role of Self-Serving Biases', *Journal of Economic Perspectives*, 11(1), 1997, 109–26, https://pubs.aeaweb.org/doi/pdfplus/10.1257/jep.11.1.109

11 A good summary is Dan Kahan's blog post, *What is Motivated Reasoning? How Does It Work?*, http://blogs.discovermagazine.com/intersection/2011/05/05/what-is-motivated-reasoning-how-does-it-work-dan-kahan-answers/#.WN5zJ_nyuUm. An excellent survey is Ziva Kunda, 'The case for motivated reasoning', *Psychological Bulletin*, 108(3), 1990, 480–98, http://dx.doi.org/10.1037/0033-2909.108.3.480

12 S. C. Kalichman, L. Eaton, C. Cherry, '"There is no proof that HIV causes AIDS": AIDS denialism beliefs among people living with HIV/AIDS', *Journal of Behavioral Medicine*, 33(6), 2010, 432–40, https://doi.org/10.1007/s10865-010-9275-7; and A. B. Hutchinson, E. B. Begley, P. Sullivan, H. A. Clark, B. C. Boyett, S. E. Kellerman, 'Conspiracy beliefs and trust in information about HIV/AIDS among minority men who have sex with men', *Journal of Acquired Immune Deficiency Syndrome*, 45(5), 15 August 2007, 603–5.

13 Tim Harford, 'Why it's too tempting to believe the Oxford study', *Financial Times,* 27 March 2020, https://www.ft.com/content/14df8908-6f47-11ea-9bca-bf503995cd6f

14 Keith E. Stanovich, Richard F. West and Maggie E. Toplak, 'Myside Bias, Rational Thinking, and Intelligence', *Current Directions in Psychological Science* 22(4), August 2013, 259–64, https://doi.org/10.1177/0963721413480174
15 Charles S. Taber and Milton Lodge, 'Motivated Skepticism in the Evaluation of Political Beliefs', *American Journal of Political Science*, 50(3), July 2006, 755–69, http://www.jstor.org/stable/3694247
16 Kevin Quealy, 'The More Education Republicans Have, the Less They Tend to Believe in Climate Change', *New York Times*, 14 November 2017, https://www.nytimes.com/interactive/2017/11/14/upshot/climate-change-by-education.html
17 Caitlin Drummond and Baruch Fischhoff, 'Individuals with greater science literacy and education have more polarized beliefs on controversial science topics', *PNAS*, 21 August 2017, http://www.pnas.org/content/early/2017/08/15/1704882114
18 Charles Lord, L. Ross and M. R. Lepper, 'Biased assimilation and attitude polarization: The effects of prior theories on subsequently considered evidence', *Journal of Personality and Social Psychology*, 37(11), 1979, 2098–2109.
19 Nicholas Epley and Thomas Gilovich, 'The Mechanics of Motivated Reasoning', *Journal of Economic Perspectives*, 30(3), 2016, 133–40, https://pubs.aeaweb.org/doi/pdfplus/10.1257/jep.30.3.133
20 Ari LeVaux, 'Climate change threatens Montana's barley farmers – and possibly your beer', Food and Environment Research Network, 13 December 2017, https://thefern.org/2017/12/climate-change-threatens-montanas-barley-farmers-possibly-beer/
21 Author correspondence with Kris De Meyer, 27 October 2018.
22 Gordon Pennycook, Ziv Epstein, Mohsen Mosleh, Antonio A. Arechar, Dean Eckles and David G. Rand. 'Understanding and Reducing the Spread of Misinformation Online.' PsyArXiv. 13 November 2019. https://doi.org/10.31234/osf.io/3n9u8; see also Oliver Burkeman, 'How to stop the spread of fake news? Pause for a moment', *Guardian,* 7 February 2020, https://www.theguardian.com/lifeandstyle/2020/feb/07/how-to-stop-spread-of-fake-news-oliver-burkeman
23 G. Pennycook and D. G. Rand, 'Lazy, not biased: Susceptibility to partisan fake news is better explained by lack of reasoning than by motivated reasoning', *Cognition*, 2018, https://doi.org/10.1016/j.cognition.2018.06.011
24 Shane Frederick, 'Cognitive Reflection and Decision Making', *Journal of Economic Perspectives*, 19(4),2005, 25–42, https://doi.org/10.1257/089533005775196732
25 Diane Wolf, *Beyond Anne Frank: Hidden Children and Postwar Families in Holland*, Berkeley: University of California Press, 2007, Table 1, citing Raul Hilberg, *The Destruction of the European Jews* (1985).

Rule Two: Ponder your personal experience

1 Muhammad Yunus interviewed by Steven Covey, http://socialbusinesspedia.com/wiki/details/248

2 Transport for London, *Travel In London: Report 11*, http://content.tfl.gov.uk/travel-in-london-report-11.pdf, figure 10.8, p.202.
3 These numbers were revealed in a freedom of information request – https://www.whatdotheyknow.com/request/journey_demand_and_service_suppl – and they are nicely summarised here: https://www.ianvisits.co.uk/blog/2016/08/05/london-tube-train-capacities/
4 Transport for London, *Travel In London: Report 4*, http://content.tfl.gov.uk/travel-in-london-report-4.pdf, p.5.
5 Author interview with Lauren Sager Weinstein and Dale Campbell of TfL, 9 July 2019.
6 Ipsos MORI, *Perils of Perception 2017*, https://www.ipsos.com/ipsos-mori/en-uk/perils-perception-2017
7 '"No link between MMR and autism", major study finds', *NHS News*, Tuesday, 5 March 2019, https://www.nhs.uk/news/medication/no-link-between-mmr-and-autism-major-study-finds/
8 'When do children usually show symptoms of autism?', *National Institute of Child Health and Clinical Development*, https://www.nichd.nih.gov/health/topics/autism/conditioninfo/symptoms-appear
9 David McRaney, 'You Are Not So Smart Episode 62: Naïve Realism', https://youarenotsosmart.com/2015/11/09/yanss-062-why-you-often-believe-people-who-see-the-world-differently-are-wrong/; and Tom Gilovich and Lee Ross, *The Wisest One in the Room*, New York: Free Press, 2016.
10 Ipsos MORI, *Perils of Perception 2017*, https://www.ipsos.com/ipsos-mori/en-uk/perils-perception-2017
11 David Dranove, Daniel Kessler, Mark McClellan and Mark Satterthwaite, 'Is More Information Better? The Effects of "Report Cards" on Health Care Providers', National Bureau of Economic Research Working Paper 8697 (2002), http://www.nber.org/papers/w8697
12 Charles Goodhart, 'Problems of Monetary Management: The U.K. Experience', in Anthony S. Courakis (ed.), *Inflation, Depression, and Economic Policy in the West*, London: Mansell, 1981, pp.111–46. The original paper was presented at a conference in 1975.
13 Donald T. Campbell, 'Assessing the impact of planned social change', *Evaluation and Program Planning*, 2(1), 1979 – an earlier version was published in 1976 and a conference paper existed in 1974.
14 Abhijit Vinayak Banerjee, Dean S. Karlan and Jonathan Zinman, 'Six randomized evaluations of microcredit: Introduction and further steps', 2015; and Rachel Meager, 'Understanding the average effect of microcredit', https://voxdev.org/topic/methods-measurement/understanding-average-effect-microcredit
15 Anna Rosling Rönnlund, 'See how the rest of the world lives, organized by income', TED 2017, anna_rosling_ronnlund_see_how_the_rest_of_the_world_lives_organized_by_income

Rule Three: Avoid premature enumeration

1 Dr Lucy Smith was interviewed by me and my colleague Richard Fenton-Smith for an episode of *More or Less* broadcast on BBC Radio 4 on 8 June 2018, https://www.bbc.co.uk/programmes/p069jd0p. The account here is based on our broadcast interview, on discussions over email, and on a phone interview I conducted with Dr Smith on 12 August 2019. Dr Smith's interviews with people who had lost a baby between twenty and twenty-four weeks of pregnancy are at https://www.healthtalk.org/20-24

2 See Merian F. MacDorman et al, 'International Comparisons of Infant Mortality and Related Factors: United States and Europe, 2010', *National Vital Statistics Reports*, 24 September 2014.

3 Denis Campbell, 'Concern at rising infant mortality rate in England and Wales', *Guardian*, 15 March 2018, https://www.theguardian.com/society/2018/mar/15/concern-at-rising-infant-mortality-rate-in-england-and-wales

4 Peter Davis et al, 'Rising infant mortality rates in England and Wales – we need to understand gestation specific mortality', *BMJ* 361, 8 May 2018, https://doi.org/10.1136/bmj.k1936

5 BBC *More or Less,* 8 April 2020, https://www.bbc.co.uk/programmes/m000h6cb

6 Author interview with Rebecca Goldin, 12 December 2017.

7 Paul J. C. Adachi and Teena Willoughby, 'The Effect of Video Game Competition and Violence on Aggressive Behavior: Which Characteristic Has the Greatest Influence?', *Psychology of Violence*, 1(4), 2011, 259–74, https://doi.org/10.1037/a0024908

8 'Immigration post-Brexit', Leave Means Leave research paper, http://www.leavemeansleave.eu/research/immigration-post-brexit-fair-flexible-forward-thinking-immigration-policy/

9 Jonathan Portes, 'Who Are You Calling Low-Skilled?', *UK in a Changing Europe*, 12 April 2017, https://ukandeu.ac.uk/who-are-you-calling-low-skilled/

10 Robert Wright, 'Brexit visa changes to hit sectors in need of low-skilled labour', *Financial Times,* 18 February 2020, https://www.ft.com/content/890e84ce-5268-11ea-90ad-25e377c0ee1f

11 https://www.theguardian.com/society/2018/nov/22/concern-over-rise-in-suicide-attempts-among-young-women

12 NHS Digital, *Mental Health of Children and Young People in England, 2017*, 22 November 2018, https://digital.nhs.uk/data-and-information/publications/statistical/mental-health-of-children-and-young-people-in-england/2017/2017

13 https://www.nhs.uk/conditions/self-harm/

14 Email correspondence with the NatCen press office, 29 November 2018.

15 Data from official sources such as the Office for National Statistics: https://www.ons.gov.uk/peoplepopulationandcommunity/birthsdeathsandmarriages/deaths/bulletins/suicidesintheunitedkingdom/2017registrations#suicide-patterns-by-age

16 https://www.theguardian.com/business/2014/jan/20/oxfam-85-richest-people-half-of-the-world
17 https://oxfamblogs.org/fp2p/anatomy-of-a-killer-fact-the-worlds-85-richest-people-own-as-much-as-poorest-3-5-billion/; and for the BBC interview with Mr Fuentes see https://www.bbc.com/news/magazine-26613682
18 The underlying data come from the *Global Wealth Report*, which is published each year by Credit Suisse. The 2013 version supplied the data for Oxfam's original 'killer fact' and it is available online here: https://publications.credit-suisse.com/tasks/render/file/?fileID=BCDB1364-A105-0560-1332EC9100FF5C83
19 'Social protection for older persons: Policy trends and statistics 2017–19', International Labour Office, Social Protection Department, Geneva, 2018; available at https://www.ilo.org/wcmsp5/groups/public/---ed_protect/---soc_sec/documents/publication/wcms_645692.pdf
20 For the UK, the Institute for Fiscal Studies *Review of Living Standards, Poverty and Inequality in the UK*. For global top incomes, the *World Inequality Report*. Another good source is *Our World In Data*. More specific references are provided in the notes below.

Rule Four: Step back and enjoy the view

1 For more reporting on this issue, listen to the 8 June 2018 episode of *More or Less*, presented by me and researched by my colleagues Richard Fenton-Smith and Richard Vadon: https://www.bbc.co.uk/programmes/p069jd0p
2 Johan Galtung and Mari Holmboe Ruge, 'The structure of foreign news: The presentation of the Congo, Cuba and Cyprus crises in four Norwegian newspapers', *Journal of Peace Research*, 2(1), 1965, 64–90.
3 Max Roser, 'Stop Saying that 2016 Was the Worst Year', *Washington Post*, 29 December 2016, https://www.washingtonpost.com/posteverything/wp/2016/12/29/stop-saying-that-2016-was-the-worst-year/?utm_term=.bad894bad69a; see also NPR's *Planet Money*, 'The Fifty Year Newspaper', 29 December 2017, https://www.npr.org/templates/transcript/transcript.php?storyId=574662798
4 C. P. Morice, J. J. Kennedy, N. A. Rayner and P. D. Jones, 'Quantifying uncertainties in global and regional temperature change using an ensemble of observational estimates: The HadCRUT4 dataset', *Journal of Geophysical Research*, 117(D8), 2012, https://doi.org/10.1029/2011JD017187, describing data from the Met Office Hadley Centre. The data are charted by and downloadable from 'Our World in Data', https://ourworldindata.org/co2-and-other-greenhouse-gas-emissions. In the 1960s, global temperatures were typically around 0.1°C below the average of 1961–90. In the twenty-first century they've typically been about 0.6°C above that average, and more recently above 0.7°C. Increase in temperatures, then, over the past fifty years, has been 0.7–0.8°C.
5 Max Roser, 'The short history of global living conditions and why it

matters that we know it', 2018, published online at OurWorldInData.org, retrieved from https://ourworldindata.org/a-history-of-global-living-conditions-in-5-charts; for Child Mortality, Roser cites data from Gapminder and the World Bank.

6 See Figure E4 in the Executive Summary of the 2018 World Inequality Report: https://wir2018.wid.world/files/download/wir2018-summary-english.pdf

7 An excellent source is the Institute for Fiscal Studies review of Living Standards, Poverty and Inequality in the UK. I've used the 2018 edition, the most recent available at the time of writing: https://www.ifs.org.uk/uploads/R145%20for%20web.pdf

8 A good summary article on inequality around the world is on the Our World in Data website, written by Joe Hasell, an authority on the subject: https://ourworldindata.org/income-inequality-since-1990

9 Author calculations, based on Natsal-3, the third National Survey of Sexual Attitudes and Lifestyles: http://timharford.com/2018/09/is-twitter-more-unequal-than-life-sex-or-happiness/

10 Michael Blastland and Andrew Dilnot, *The Tiger That Isn't*, London: Profile Books, 2008.

11 Andrew C. A. Elliott, *Is That a Big Number?*, Oxford: Oxford University Press, 2018.

12 Tali Sharot, 'The Optimism Bias', TED Talk, 2012: https://www.ted.com/talks/tali_sharot_the_optimism_bias/transcript#t-18026

13 Daniel Kahneman, *Thinking, Fast and Slow*, New York: Farrar, Straus and Giroux, 2010.

14 Ross A. Miller & Karen Albert, 'If It Leads, It Bleeds (and If It Bleeds, It Leads): Media Coverage and Fatalities in Militarized Interstate Disputes' *Political Communication* 2015, 32(1), 61–82, https://doi.org/10.1080/10584609.2014.880976; Barbara Combs & Paul Slovic, 'Newspaper Coverage of Causes of Death', *Journalism Quarterly*, 56(4), 837–43, 849.

15 https://www.cdc.gov/tobacco/data_statistics/fact_sheets/fast_facts/ – there are 1300 deaths a day from smoking-related diseases, about 40,000 a month; almost 3000 people were killed by the 11 September attacks.

16 https://www.ted.com/talks/the_ted_interview_steven_pinker_on_why_our_pessimism_about_the_world_is_wrong/transcript?language=en

17 Steven Pinker mentions in the endnotes of *Enlightenment Now* (New York: Penguin, 2018),that this correspondence took place in 1982.

18 Quoted in the *Guardian*, 12 May 2015, https://www.theguardian.com/society/2015/may/12/stroke-association-warns-of-alarming-rise-in-number-of-victims; see also *More or Less*, 17 May 2015, with the analysis of this claim: https://www.bbc.co.uk/programmes/b05tpz78

19 Oxfam press release, 22 September 2016, http://oxfamapps.org/media/ppdwr

20 A useful survey of various relevant graphs is Max Roser and Mohamed Nagdy, 'Optimism & Pessimism', 2018, published online at

OurWorldInData.org, retrieved from https://ourworldindata.org/optimism-pessimism – particularly Section I.1 with graphs from Eurobarometer and Ipsos MORI.

21 Martyn Lewis, 'Not My Idea of Good News', *Independent*, 26 April 1993, https://www.independent.co.uk/voices/not-my-idea-of-good-news-at-the-end-of-a-week-of-horrifying-events-martyn-lewis-bbc-presenter-argues-1457539.html

22 Max Roser, https://ourworldindata.org/a-history-of-global-living-conditions-in-5-charts – underlying data from the World Bank and from F. Bourguignon and C. Morrisson, 'Inequality Among World Citizens: 1820–1992', *American Economic Review*, 92(4), 2002, 727–48. In 1993 there were 1.94 billion people living in extreme poverty; by 2015 that had fallen to 0.7 billion (705.55 million). The rate of improvement averages 153,600 a day, although of course we have no way of measuring the daily rate as it fluctuates.

23 Samantha Vanderslott, Bernadeta Dadonaite and Max Roser, ' Vaccination', 2020. Published online at OurWorldInData.org. Retrieved from: https://ourworldindata.org/vaccination

24 Anna Rosling Rönnlund, Hans Rosling and Ola Rosling, *Factfulness*, London: Sceptre, 2018.

25 Gillian Tett, 'Silos and Silences', *Banque de France Financial Stability Review* No. 14 – Derivatives – Financial innovation and stability, July 2010, https://core.ac.uk/download/pdf/6612179.pdf

26 Rolf Dobelli, 'News is bad for you – and giving up reading it will make you happier', *Guardian*, 12 April 2013, https://www.theguardian.com/media/2013/apr/12/news-is-bad-rolf-dobelli

27 Nassim Nicholas Taleb, *The Bed of Procrustes*, London: Penguin Books, 2010.

28 Bill Hanage, Mark Lipsitch, 'How to Report on the COVID-19 Outbreak Responsibly', *Scientific American*, 23 February 2020, https://blogs.scientificamerican.com/observations/how-to-report-on-the-covid-19-outbreak-responsibly/

Rule Five: Get the back story

1 Sheena Iyengar and Mark Lepper, 'When Choice is Demotivating: Can One Desire Too Much of a Good Thing?', *Journal of Personality and Social Psychology*, 79, 2000.

2 Author interview with Benjamin Scheibehenne, October 2009. (I'd like to claim I was ahead of the curve on this one.)

3 B. Scheibehenne, R. Greifeneder and P. M. Todd, 'Can There Ever Be Too Many Options? A Meta-Analytic Review of Choice Overload', *Journal of Consumer Research*, 37, 2010, 409–25, http://scheibehenne.de/ScheibehenneGreifenederTodd2010.pdf

4 'Ten Kickstarter Products that Raised the Most Money': https://www.marketwatch.com/story/10-kickstarter-products-that-raised-the-most-money-2017-06-22-10883052

5 The story is well told in Jordan Ellenberg's book *How Not to Be Wrong* (New York: Penguin Press, 2014), with the relevant extract here: https://medium.com/@penguinpress/an-excerpt-from-how-not-to-be-wrong-by-jordan-ellenberg-664e708cfc3d

6 A technical summary (along with some grumbling about how the story has been exaggerated) is in Bill Casselman, 'The Legend of Abraham Wald', American Mathematical Society, http://www.ams.org/publicoutreach/feature-column/fc-2016-06

7 An excellent account of the controversy is Daniel Engber, 'Daryl Bem Proved ESP Is Real Which Means Science Is Broken', *Slate*, 17 May 2017, https://slate.com/health-and-science/2017/06/daryl-bem-proved-esp-is-real-showed-science-is-broken.html

8 Chris French, 'Precognition studies and the curse of the failed replications', *Guardian*, 15 March 2012, https://www.theguardian.com/science/2012/mar/15/precognition-studies-curse-failed-replications

9 Nosek was speaking to the *Planet Money* podcast, episode 677: https://www.npr.org/sections/money/2018/03/07/591213302/episode-677-the-experiment-experiment

10 Brian Nosek has given useful interviews to several podcasts, including *You Are Not So Smart* (episode 100), https://youarenotsosmart.com/2017/07/19/yanss-100-the-replication-crisis/; *Planet Money* (episode 677), https://www.npr.org/sections/money/2018/03/07/591213302/episode-677-the-experiment-experiment; *EconTalk* (16 November 2015), http://www.econtalk.org/brian-nosek-on-the-reproducibility-project/; *The Hidden Brain* (episode 32), https://www.npr.org/templates/transcript/transcript.php?storyId=477921050; as well as BBC *Analysis*, 'The Replication Crisis', 12 November 2018, https://www.bbc.co.uk/programmes/m00013p9

11 This figure of thirty-nine is based on the subjective opinion of the replicating researchers. Did their results basically back up the original study, or not? That's a judgement call. An alternative metric is to ask how many of the replication studies produced results that passed the standard (but rather problematic) hurdle of 'statistical significance'. Only thirty-six did; ninety-seven of the original studies had cleared that hurdle. See 'Estimating the reproducibility of psychological science' by the Open Science Collaboration, published in *Science*, 28 August 2015, 349(6251), https://doi.org/10.1126/science.aac4716.

12 Brief film on YouTube here: https://www.youtube.com/watch?v=n1SJ-Tn3bcQ

13 *Planet Money*, episode 677: https://www.npr.org/sections/money/2018/03/07/591213302/episode-677-the-experiment-experiment

14 F. J. Anscombe, 'Fixed-Sample-Size Analysis of Sequential Observations', *Biometrics*, 10(1), 1954, 89–100, www.jstor.org/stable/3001665; and Andrew Gelman, *Statistical Inference, Modelling and Social Science*, blog post 2 May 2018, https://statmodeling.stat.columbia.edu/2018/05/02/continuously-increased-number-animals-statistical-significance-reached-support-conclusions-think-not-bad-actually/

15 David J. Hand, *Dark Data*, Princeton: Princeton University Press, 2020.
16 Andrew Gelman and Eric Loken, 'The garden of forking paths: Why multiple comparisons can be a problem, even when there is no "fishing expedition" or "p-hacking" and the research hypothesis was posited ahead of time', working paper, 14 November 2013, http://www.stat.columbia.edu/~gelman/research/unpublished/p_hacking.pdf
17 J. P. Simmons, L. D. Nelson & U. Simonsohn, 'False-Positive Psychology: Undisclosed Flexibility in Data Collection and Analysis Allows Presenting Anything as Significant', *Psychological Science*, 22(11), 2011, 1359–66, https://doi.org/10.1177/0956797611417632
18 Kai Kupferschmidt, 'More and more scientists are preregistering their studies. Should you?', *Science*, 21 September 2018.
19 Anjana Ahuja, 'Scientists strike back against statistical tyranny', *Financial Times*, 27 March 2019, https://www.ft.com/content/36f9374c-5075-11e9-8f44-fe4a86c48b33
20 Darrell Huff, *How to Lie with Statistics*, New York: W. W. Norton, 1993, p.40.
21 John Ioannidis, 'Why Most Published Research Findings Are False', *PLoS Medicine*, 2(8), August 2005, e124, https://doi.org/10.1371/journal.pmed.0020124
22 R. F. Baumeister, E. Bratslavsky, M. Muraven and D. M. Tice, 'Ego depletion: Is the active self a limited resource?', *Journal of Personality and Social Psychology*, 74(5), 1998, 1252–65, http://dx.doi.org/10.1037/0022-3514.74.5.1252; and 'The End of Ego Depletion Theory?', *Neuroskeptic* blog, 31 July 2016, http://blogs.discovermagazine.com/neuroskeptic/2016/07/31/end-of-ego-depletion/#.XGGyflz7SUk
23 Amy Cuddy, 'Your Body Language May Shape Who You Are', TED Talk, 2012, https://www.ted.com/talks/amy_cuddy_your_body_language_shapes_who_you_are/transcript?language=en
24 Kahneman, *Thinking, Fast and Slow*, pp.53–7.
25 Ed Yong, 'Nobel laureate challenges psychologists to clean up their act', *Nature News*, 3 October 2012, https://www.nature.com/news/nobel-laureate-challenges-psychologists-to-clean-up-their-act-1.11535
26 Ben Goldacre, 'Backwards Step on Looking into the Future', *Guardian*, 23 April 2011, https://www.theguardian.com/commentisfree/2011/apr/23/ben-goldacre-bad-science
27 Robin Wrigglesworth, 'How a herd of cows trampled on human stockpickers', *Financial Times*, 21 January 2020, https://www.ft.com/content/563d61dc-3b70-11ea-a01a-bae547046735?
28 Burton Malkiel, 'Returns from Investing in Equity Funds', working paper, Princeton University, 1994.
29 Eric Balchunas, 'How the Vanguard Effect adds up to $1 trillion', Bloomberg.com, 30 August 2016, https://www.bloomberg.com/opinion/articles/2016-08-30/how-much-has-vanguard-saved-investors-try-1-trillion
30 For an accessible overview, see Ben Goldacre, 'What doctors

don't know about the drugs they prescribe', TED Talk, 2012, https://www.ted.com/talks/ben_goldacre_what_doctors_don_t_know_about_the_drugs_they_prescribe/footnotes?language=en

31 Erick Turner et al, 'Selective Publication of Antidepressant Trials and Its Influence on Apparent Efficacy', *New England Journal of Medicine*, 17 January 2008, https://www.nejm.org/doi/full/10.1056/NEJMsa065779

32 Ben Goldacre, 'Transparency, Beyond Publication Bias', talk given to the International Journal of Epidemiology Conference, 2016; available at https://www.badscience.net/2016/10/transparency-beyond-publication-bias-a-video-of-my-super-speedy-talk-at-ije/

33 Ben Goldacre, Henry Drysdale, Aaron Dale, Ioan Milosevic, Eirion Slade, Philip Hartley, Cicely Marston, Anna Powell-Smith, Carl Heneghan and Kamal R. Mahtani, 'COMPare: a prospective cohort study correcting and monitoring 58 misreported trials in real time', *Trials*, 20(118), 2019, https://doi.org/10.1186/s13063-019-3173-2.

34 Goldacre, 'Transparency, Beyond Publication Bias' , https://www.badscience.net/2016/10/transparency-beyond-publication-bias-a-video-of-my-super-speedy-talk-at-ije/

35 Amy Sippett, 'Does the Backfire Effect exist?', *Full Fact*, 20 March 2019, https://fullfact.org/blog/2019/mar/does-backfire-effect-exist/; Brendan Nyhan tweet, 20 March 2019, https://twitter.com/BrendanNyhan/status/1108377656414879744

36 Author interview with Richard Thaler, 17 July 2019.

37 BBC *Analysis*, 'The Replication Crisis', 12 November 2018, https://www.bbc.co.uk/programmes/m00013p9

38 Antonio Granado, 'Slaves to journals, serfs to the web: The use of the internet in newsgathering among European science journalists', *Journalism*, 12(7), 2011, 794–813.

39 A. L. Cochrane, 'Sickness in Salonica: My first, worst, and most successful clinical trial', *British Medical Journal (Clin Res Ed)*, 289(6460), 1984, 1726–7, https://doi.org/10.1136/bmj.289.6460.1726

40 'A Brief History of Cochrane', https://community.cochrane.org/handbook-sri/chapter-1-introduction/11-cochrane/112-brief-history-cochrane

41 https://www.webmd.com/urinary-incontinence-oab/news/20180522/yoga-may-be-right-move-versus-urinary-incontinence#1

42 https://www.dailymail.co.uk/health/article-2626209/Could-yoga-cure-INCONTINENCE-Exercise-strengthens-pelvic-floor-muscles-reducing-leakage.html

43 https://www.hcd.com/incontinence/yoga-incontinence/

44 https://www.ncbi.nlm.nih.gov/pmc/articles/PMC4310548/

45 L. S. Wieland, N. Shrestha, Z. S. Lassi, S. Panda, D. Chiaramonte and N. Skoetz, 'Yoga for treating urinary incontinence in women', *Cochrane Database of Systematic Reviews 2019*, 2, Art. No.: CD012668, https://doi.org/10.1002/14651858.CD012668.pub2

Rule Six: Ask who is missing

1 R. Bond and P. B. Smith, 'Culture and conformity: A meta-analysis of studies using Asch's (1952b, 1956) line judgment task', *Psychological Bulletin*, 119(1), 1996, 111–37, http://dx.doi.org/10.1037/0033-2909.119.1.111

2 Tim Harford, 'The Truth About Our Norm-Core', *Financial Times*, 12 June 2015, http://timharford.com/2015/06/the-truth-about-our-norm-core/

3 Bond and Smith, 'Culture and conformity'; and Natalie Frier, Colin Fisher, Cindy Firman and Zachary Bigaouette, 'The Effects of Group Conformity Based on Sex', 2016, Celebrating Scholarship & Creativity Day, Paper 83, http://digitalcommons.csbsju.edu/elce_cscday/83

4 Tim Harford, 'Trump, Brexit and How Politics Loses the Capacity to Shock', *Financial Times*, 16 November 2018, https://www.ft.com/content/b730c95c-e82e-11e8-8a85-04b8afea6ea3

5 Caroline Criado Perez, *Invisible Women*, London: Chatto and Windus, 2019; the interview was broadcast on BBC Radio 4 on 17 May 2019 and is available on the *More or Less* website: https://www.bbc.co.uk/programmes/m00050rd

6 Peter Hofland, 'Reversal of Fortune', *Onco'Zine*, 30 November 2013, https://oncozine.com/reversal-of-fortune-how-a-vilified-drug-became-a-life-saving-agent-in-the-war-against-cancer/

7 R. Dmitrovic, A. R. Kunselman, R. S. Legro, 'Sildenafil citrate in the treatment of pain in primary dysmenorrhea: a randomized controlled trial', *Human Reproduction*, 28(11), November 2013, 2958–65, https://doi.org/10.1093/humrep/det324

8 BBC *More or Less*, 31 March 2020, https://www.bbc.co.uk/sounds/play/m000h7st

9 Mayra Buvinic and Ruth Levine, 'Closing the gender data gap', *Significance*, 8 April 2016, https://doi.org/10.1111/j.1740-9713.2016.00899.x; and Charlotte McDonald, 'Is There a Sexist Data Crisis?', BBC News, 18 May 2016, https://www.bbc.co.uk/news/magazine-36314061

10 Shelly Lundberg, Robert Pollak and Terence J. Wales, 'Do Husbands and Wives Pool Their Resources? Evidence from the United Kingdom Child Benefit', 32(3), 1997, 463–80, https://econpapers.repec.org/article/uwpjhriss/v_3a32_3ay_3a1997_3ai_3a3_3ap_3a463-480.htm

11 Buvinic and Levine, 'Closing the gender data gap', https://doi.org/10.1111/j.1740-9713.2016.00899.x

12 Suzannah Brecknell, 'Interview: Full Fact's Will Moy on lobbyist "nonsense", official corrections and why we know more about golf than crime stats', *Civil Service World*, 5 May 2016, https://www.civilserviceworld.com/articles/interview/interview-full-fact%E2%80%99s-will-moy-lobbyist-%E2%80%9Cnonsense%E2%80%9D-official-corrections-and-why

13 Maurice C. Bryson, 'The Literary Digest Poll: Making of a Statistical Myth', *American Statistician*, 30(4), 1976, 184–5, https://doi.org/10.1080/00031305.1976.10479173; and Peverill Squire, 'Why the 1936

Literary Digest Poll Failed', *Public Opinion Quarterly*, 52(1), 1988, 125–33, www.jstor.org/stable/2749114

14 P. Whiteley, 'Why Did the Polls Get It Wrong in the 2015 General Election? Evaluating the Inquiry into Pre-Election Polls', *Political Quarterly*, 87, 2016, 437–42, https://doi.org/10.1111/1467-923X.12274

15 John Curtice, 'Revealed: Why the Polls Got It So Wrong in the British General Election', *The Conversation*, 14 January 2016, https://theconversation.com/revealed-why-the-polls-got-it-so-wrong-in-the-british-general-election-53138

16 Nate Cohn, 'A 2016 Review: Why Key State Polls Were Wrong About Trump', *New York Times*, 31 May 2017, https://www.nytimes.com/2017/05/31/upshot/a-2016-review-why-key-state-polls-were-wrong-about-trump.html; and Andrew Mercer, Claudia Deane and Kyley McGeeney, 'Why 2016 election polls missed their mark', Pew Research Fact Tank blog, 9 November 2015, http://www.pewresearch.org/fact-tank/2016/11/09/why-2016-election-polls-missed-their-mark/

17 https://www.ons.gov.uk/peoplepopulationandcommunity/populationandmigration/populationestimates/methodologies/2011censusstatisticsforenglandandwalesmarch2011qmi

18 Author interview with Viktor Mayer-Schönberger, March 2014.

19 Pew Research Center Social Media Factsheet, research conducted January 2018, https://www.pewinternet.org/fact-sheet/social-media/

20 Kate Crawford, 'The Hidden Biases in Big Data', *Harvard Business Review*, 1 April 2013, https://hbr.org/2013/04/the-hidden-biases-in-big-data

21 Leon Kelion, 'Coronavirus: Covid-19 detecting apps face teething problems', BBC News, 8 April 2020, https://www.bbc.co.uk/news/technology-52215290

22 Kate Crawford, 'Artificial Intelligence's White Guy Problem', *New York Times*, 25 June 2016, https://www.nytimes.com/2016/06/26/opinion/sunday/artificial-intelligences-white-guy-problem.html

Rule Seven: Demand transparency when the computer says 'no'

1 Jeremy Ginsberg, Matthew H. Mohebbi, Rajan S. Patel, Lynnette Brammer, Mark S. Smolinski, Larry Brilliant, 'Detecting influenza epidemics using search engine query data', *Nature*, 457 (7232), 19 February 2009, 1012–14, https://doi.org/10.1038/nature07634

2 Parts of this chapter are closely based on my *Financial Times* magazine article 'Big Data: Are We Making a Big Mistake?' (*FT*, 28 March 2014, https://www.ft.com/content/21a6e7d8-b479-11e3-a09a-00144feabdc0). I interviewed David Hand, Kaiser Fung, Viktor Mayer-Schönberger, David Spiegelhalter and Patrick Wolfe in early 2014 for the piece.

3 David Lazer and Ryan Kennedy, 'What We Can Learn from the Epic Failure of Google Flu Trends', *Wired*, https://www.wired.com/2015/10/can-learn-epic-failure-google-flu-trends/; and Declan Butler, 'What

Google Flu Got Wrong', *Nature*, https://www.nature.com/news/when-google-got-flu-wrong-1.12413

4 https://www.google.org/flutrends/about/

5 D. Lazer, R. Kennedy, G. King and A. Vespignani, 'The Parable of Google Flu: Traps in Big Data Analysis', *Science* 343(6176), March 2014, 1203–5.

6 S. Cook, C. Conrad, A. L. Fowlkes, M. H. Mohebbi, 'Assessing Google Flu Trends Performance in the United States during the 2009 Influenza Virus A (H1N1) Pandemic', *PLoS ONE* 6(8), 2011, e23610, https://doi.org/10.1371/journal.pone.0023610

7 Janelle Shane, *You Look Like a Thing and I Love You*, New York: Little, Brown, 2019.

8 For comprehensive reporting of this affair, see the *Observer/Guardian* website: https://www.theguardian.com/news/series/cambridge-analytica-files

9 Charles Duhigg, 'How Companies Learn Your Secrets', *New York Times* magazine, 19 February 2012, https://www.nytimes.com/2012/02/19/magazine/shopping-habits.html

10 Hannah Fry, *Hello World: Being Human in ihe Age of Computers*, London: W. W. Norton, 2018.

11 Cathy O'Neil, *Weapons of Math Destruction*, London: Allen Lane, 2016.

12 *Freakonomics* radio episode 268: Bad Medicine Pt 1, 16 August 2017, http://freakonomics.com/podcast/bad-medicine-part-1-story-rebroadcast/

13 P. A. Mackowiak, S. S. Wasserman, M. M. Levine, 'A Critical Appraisal of 98.6°F, the Upper Limit of the Normal Body Temperature, and Other Legacies of Carl Reinhold August Wunderlich', *JAMA*, 268(12), 1992, 1578–80, https://doi.org/10.1001/jama.1992.03490120092034

14 Jeffrey Dastin, 'Amazon scraps secret AI recruiting tool that showed bias against women', Reuters, 10 October 2018, https://www.reuters.com/article/us-amazon-com-jobs-automation-insight/amazon-scraps-secret-ai-recruiting-tool-that-showed-bias-against-women-idUSKCN1MK08G

15 Gerd Gigerenzer and Stephanie Kurzenhaeuser, 'Fast and frugal heuristics in medical decision making', *Science and Medicine in Dialogue: Thinking through particulars and universals*, 2005, 3–15.

16 Paul Meehl, *Clinical vs. Statistical Prediction*, Minneapolis: University of Minnesota Press, 1954.

17 Fry, *Hello World.*

18 Mandeep K. Dhami and Peter Ayton, 'Bailing and jailing the fast and frugal way', *Journal of Behavioral Decision Making*, 14(2), 2001, https://doi.org/10.1002/bdm.371

19 Jon Kleinberg, Himabindu Lakkaraju, Jure Leskovec, Jens Ludwig, Sendhil Mullainathan, 'Human Decisions and Machine Predictions', *Quarterly Journal of Economics*, 133(1), February 2018, 237–93, https://doi.org/10.1093/qje/qjx032; see also Cass R. Sunstein, 'Algorithms, Correcting Biases', working paper, 12 December 2018.

20 David Jackson and Gary Marx, 'Data mining program designed to predict child abuse proves unreliable, DCFS says', *Chicago Tribune*, 6 December

2017; and Dan Hurley, 'Can an Algorithm Tell When Kids Are in Danger?', *New York Times* magazine, 2 January 2018, https://www.nytimes.com/2018/01/02/magazine/can-an-algorithm-tell-when-kids-are-in-danger.html

21 Hurley, 'Can an Algorithm Tell When Kids Are in Danger?'

22 Andrew Gelman, 'Flaws in stupid horrible algorithm revealed because it made numerical predictions', *Statistical Modeling, Causal Inference, and Social Science* blog, 3 July 2018, https://statmodeling.stat.columbia.edu/2018/07/03/flaws-stupid-horrible-algorithm-revealed-made-numerical-predictions/

23 Sabine Hossenfelder, 'Blaise Pascal, Florin Périer, and the Puy de Dôme experiment', http://backreaction.blogspot.com/2007/11/blaise-pascal-florin-p-and-puy-de-d.html; and David Wootton, *The Invention of Science: A New History of the Scientific Revolution*, London: Allen Lane, 2015, Chapter 8.

24 See, for example, Louis Trenchard More, 'Boyle as Alchemist', *Journal of the History of Ideas*, 2(1), January 1941, 61–76; and 'The Strange, Secret History of Isaac Newton's Papers', a Q&A with Sarah Dry, https://www.wired.com/2014/05/newton-papers-q-and-a/

25 Wootton, *The Invention of Science*, p.340.

26 James Burke, *Connections*, Boston: Little, Brown, 1978; reprint with new introduction 1995, p.74.

27 Wootton, *The Invention of Science*, p.357.

28 https://www.propublica.org/article/how-we-analyzed-the-compas-recidivism-algorithm

29 Sam Corbett-Davies, Emma Pierson, Avi Feller, Sharad Goel, Aziz Huq, 'Algorithmic decision making and the cost of fairness', arXiv:1701.08230; and Sam Corbett-Davies, Emma Pierson, Avi Feller and Sharad Goel, 'A computer program used for bail and sentencing decisions was labeled biased against blacks. It's actually not that clear', *Washington Post*, 17 October 2016, https://www.washingtonpost.com/news/monkey-cage/wp/2016/10/17/can-an-algorithm-be-racist-our-analysis-is-more-cautious-than-propublicas/

30 Ed Yong, 'A Popular Algorithm Is No Better at Predicting Crimes than Random People', *The Atlantic*, 17 January 2018, https://www.theatlantic.com/technology/archive/2018/01/equivant-compas-algorithm/550646/

31 Ibid.

32 Julia Dressel and Hany Farid, 'The Accuracy, Fairness and Limits of Predicting Recidivism', *Science Advances 2018*, http://advances.sciencemag.org/content/4/1/eaao5580

33 Onora O'Neill's Reith Lectures on Trust (http://www.bbc.co.uk/radio4/reith2002/) and her TED talk (https://www.ted.com/speakers/onora_o_neill) are both well worth listening to. Themes of intelligent openness are explored in depth in the Royal Society report 'Science as an Open Enterprise', 2012, of which O'Neill was an author. In his book *The Art of Statistics* (London: Penguin, 2019), David Spiegelhalter shows how O'Neill's principles can be applied to evaluating algorithms.

34 Email interview with Cathy O'Neil, 29 August 2019.

35 Jack Nicas, 'How YouTube Drives Viewers to the Internet's Darkest Corners', *Wall Street Journal*, 7 February 2018, https://www.wsj.com/articles/how-youtube-drives-viewers-to-the-internets-darkest-corners-1518020478; and Zeynep Tufekci, 'YouTube, the Great Radicalizer', *New York Times*, 10 March 2018, https://www.nytimes.com/2018/03/10/opinion/sunday/youtube-politics-radical.html. But see in contrast Mark Ledwich and Anna Zaitsev, 'Algorithmic Extremism: Examining YouTube's Rabbit Hole of Radicalization', https://arxiv.org/abs/1912.11211

36 Ryan Singal, 'Netflix spilled your Brokeback Mountain secret, Lawsuit Claims', *Wired*, 17 December 2009, https://www.wired.com/2009/12/netflix-privacy-lawsuit/; and Blake Hallinan and Ted Striphas, 'Recommended for you: the Netflix Prize and the production of algorithmic culture', *New Media and Society*, 2016, https://journals.sagepub.com/doi/pdf/10.1177/1461444814538646

Rule Eight: Don't take statistical bedrock for granted

1 This is a translation of a Danish TV interview, discussed here: https://www.thelocal.se/20150905/hans-rosling-you-cant-trust-the-media

2 Laura Smith, 'In 1974, a stripper known as the "Tidal Basin Bombshell" took down the most powerful man in Washington', *Timeline*, 18 September 2017, https://timeline.com/wilbur-mills-tidal-basin-3c29a8b47ad1; Stephen Green and Margot Hornblower, 'Mills Admits Being Present During Tidal Basin Scuffle', *Washington Post*, 11 October 1974.

3 'The Stripper and the Congressman: Fanne Foxe's Story', The Rialto Report Podcast, Episode 82, https://www.therialtoreport.com/2018/07/15/fanne-foxe/

4 Alice M. Rivlin, 'The 40th Anniversary of the Congressional Budget Office', *Brookings: On the Record*, 2 March 2015, https://www.brookings.edu/on-the-record/40th-anniversary-of-the-congressional-budget-office/

5 Philip Joyce, 'The Congressional Budget Office at Middle Age', *Hutchins Center at Brookings*, Working Paper #9, 17 February 2015.

6 Quoted in Nancy D. Kates, *Starting from Scratch: Alice Rivlin and the Congressional Budget Office*, Cambridge: John F. Kennedy School of Government, Harvard University, 1989.

7 Elaine Povich, 'Alice Rivlin, budget maestro who "helped save Washington" in fiscal crisis, dies at 88', *Washington Post*, 14 May 2019, https://www.washingtonpost.com/local/obituaries/alice-rivlin-budget-maestro-who-helped-save-washington-in-fiscal-crisis-dies-at-88/2019/05/14/c141c996-0ff9-11e7-ab07-07d9f521f6b5_story.html

8 Andrew Prokop, 'The Congressional Budget Office, explained', Vox, 26 June 2017, https://www.vox.com/policy-and-politics/2017/3/13/14860856/congressional-budget-office-cbo-explained

9 John Frendreis and Raymond Tatalovich, 'Accuracy and Bias in Macroeconomic Forecasting by the Administration, the CBO, and the Federal Reserve Board', *Polity* 32(4), 2000, 623–32, accessed 17 January

2020, https://doi.org/10.2307/3235295; Holly Battelle, *CBO's Economic Forecasting Record*, Washington DC: Congressional Budget Office, 2010; Committee for a Responsible Federal Budget, 'Hindsight is 2020: A look back at CBO's economic forecasting', January 2013, https://www.crfb.org/blogs/hindsight-2020-look-back-cbos-economic-forecasting

10 *Forecast Evaluation Report 2019*, Office for Budget Responsibility, December 2019, https://obr.uk/docs/dlm_uploads/Forecast_evaluation_report_December_2019-1.pdf

11 Malcolm Bull, 'Can the Poor Think?', *London Review of Books*, 41(13), 4 July 2019.

12 Bourree Lam, 'After a Good Jobs Report, Trump Now Believes Economic Data', *The Atlantic*, 10 March 2017, https://www.theatlantic.com/business/archive/2017/03/trump-spicer-jobs-report/519273/

13 Esther King, 'Germany records lowest crime rate since 1992', *Politico*, 8 May 2017, https://www.politico.eu/article/germany-crime-rate-lowest-since-1992/

14 For discussion and the full Trump tweets, see Matthew Yglesias, 'Trump just tweeted that "crime in Germany is way up." It's actually at its lowest level since 1992', Vox, 18 June 2018; and Christopher F. Schuetze and Michael Wolgelenter, 'Fact Check: Trump's False and Misleading Claims about Germany's Crime and Immigration', *New York Times*, 18 June 2018.

15 Diane Coyle, *GDP: A Brief But Affectionate History*, Oxford: Princeton University Press, 2014, pp.3–4.

16 'Report on Greek government deficit and debt statistics', European Commission, 8 January 2010.

17 Beat Balzli, 'Greek Debt Crisis: How Goldman Sachs Helped Greece to Mask its True Debt', *Der Spiegel*, 8 February 2010, https://www.spiegel.de/international/europe/greek-debt-crisis-how-goldman-sachs-helped-greece-to-mask-its-true-debt-a-676634.html

18 The International Statistical Institute has a chronological account of the sorry tale – last updated by G. O'Hanlon and H. Snorrason, July 2018: https://isi-web.org/images/news/2018-07_Court-proceedings-against-Andreas-Georgiou.pdf

19 'Commendation of Andreas Georgiou' – Press Release: International Statistical Association, 18 September 2018, https://www.isi-web.org/images/2018/Press%20release%20Commendation%20for%20Andreas%20Georgiou%20Aug%202018.pdf

20 R. Langkjær-Bain, 'Trials of a statistician', *Significance*, 14, 2017, 14–19, https://doi.org/10.1111/j.1740-9713.2017.01052.x; 'An Augean Stable', *The Economist*, 13 February 2016, https://www.economist.com/the-americas/2016/02/13/an-augean-stable; 'The Price of Cooking the Books', *The Economist*, 25 February 2012, https://www.economist.com/the-americas/2012/02/25/the-price-of-cooking-the-books

21 Langkjær-Bain, 'Trials of a statistician'.

22 Author interview with Denise Lievesley, 2 July 2018.

23 'Tanzania law punishing critics of statistics "deeply concerning": World

Bank', Reuters, 3 October 2018, https://www.reuters.com/article/us-tanzania-worldbank/tanzania-law-punishing-critics-of-statistics-deeply-concerning-world-bank-idUSKCN1MD17P

24 Amy Kamzin, 'Dodgy data makes it hard to judge Modi's job promises', *Financial Times*, 8 October 2018, https://www.ft.com/content/1a008ebe-cad4-11e8-9fe5-24ad351828ab

25 Steven Chase and Tavia Grant, 'Statistics Canada chief falls on sword over census', *Globe and Mail*, 21 July 2010, https://www.theglobeandmail.com/news/politics/statistics-canada-chief-falls-on-sword-over-census/article1320915/

26 Langkjær-Bain, 'Trials of a statistician'.

27 Nicole Acevedo, 'Puerto Rico faces lawsuits over hurricane death count data', NBC News, 1 June 2018; and Joshua Barajas, 'Hurricane Maria's official death toll is 46 times higher than it was almost a year ago. Here's why', PBS Newshour, 30 August 2018, https://www.pbs.org/newshour/nation/hurricane-marias-official-death-toll-is-46-times-higher-than-it-was-almost-a-year-ago-heres-why

28 '2011 Census Benefits Evaluation Report', https://www.ons.gov.uk/census/2011census/2011censusbenefits/2011censusbenefitsevaluationreport#unquantified-benefits; Ian Cope, 'The Value of Census Statistics', https://www.ukdataservice.ac.uk/media/455474/cope.pdf

29 Carl Bakker, *Valuing the Census*, 2014, https://www.stats.govt.nz/assets/Research/Valuing-the-Census/valuing-the-census.pdf

30 Mónica I. Feliú-Mójer, 'Why Is Puerto Rico Dismantling Its Institute of Statistics?', *Scientific American: Voices*, 1 February 2018.

31 https://www.cbo.gov/publication/54965

32 Ellen Hughes-Cromwick and Julia Coronado, 'The Value of US Government Data to US Business Decisions', *Journal of Economic Perspectives*, 33(1), 2019, 131–46, https://doi.org/10.1257/jep.33.1.131.

33 Milton and Rose Friedman, *Two Lucky People* (1998), quoted in Neil Monnery, 'Hong Kong's postwar transformation shows how fewer data can sometimes boost growth', https://blogs.lse.ac.uk/businessreview/2017/06/30/hong-kongs-postwar-transformation-shows-how-fewer-data-can-sometimes-boost-growth/

34 James C. Scott, *Seeing Like a State: How Certain Schemes to Improve the Human Condition Have Failed*, New Haven: Yale University Press, 1998.

35 Perry Link, 'China: From Famine to Oslo', *New York Review of Books*, 13 January 2011.

36 For a discussion of the death toll under Stalin, see Timothy Snyder, 'Hitler vs. Stalin: Who Killed More?', *New York Review of Books*, 10 March 2011 – a more sensitively written piece than the title suggests. For more on the 1937 census, see Daniel Sandford, 'In Moscow, history is everywhere', BBC News, 2 November 2012; and Catherine Merridale, 'The 1937 Census and the Limits of Stalinist Rule', *Historical Journal*, 39(1), 1996, and 'Called to Account', *The Economist*, 3 September 2016, https://www.economist.com/finance-and-economics/2016/09/03/called-to-account

37 Merridale, 'The 1937 Census and the Limits of Stalinist Rule'.

38 Adam Tooze, *Statistics and the German State, 1900-1945*, Cambridge: Cambridge University Press, 2001, p.257.

39 Author interview with Denise Lievesley, 11 March 2019.

40 Hetan Shah, 'How to save statistics from the threat of populism', *Financial Times*, 21 October 2018, https://www.ft.com/content/ca491f18-d383-11e8-9a3c-5d5eac8f1ab4

41 Nicholas Eberstadt, Ryan Nunn, Diane Whitmore Schanzenbach, Michael R. Strain, '"In Order That They Might Rest Their Arguments on Facts": The Vital Role of Government-Collected Data', AEI/Hamilton Project report, March 2017.

42 For more on the Rayner Review, see G. Hoinville and T. M. F. Smith, 'The Rayner Review of Government Statistical Services', *Journal of the Royal Statistical Society*, Series A (General) 145(2),1982, 195–207, https://doi.org/10.2307/2981534; and John Kay, 'A Better Way to Restore Faith in Official Statistics', 25 July 2006, https://www.johnkay.com/2006/07/25/a-better-way-to-restore-faith-in-official-statistics/

43 Hughes-Cromwick and Coronado, 'The Value of US Government Data to US Business Decisions', https://doi.org/10.1257/jep.33.1.131

44 Jackie Mansky, 'W.E.B. Du Bois' Visionary Infographics Come Together for the First Time in Full Color', *Smithsonian Magazine*, 15 November 2018, https://www.smithsonianmag.com/history/first-time-together-and-color-book-displays-web-du-bois-visionary-infographics-180970826/; and Mona Chalabi, 'WEB Du Bois: retracing his attempt to challenge racism with data', *Guardian*, 14 February 2017, https://www.theguardian.com/world/2017/feb/14/web-du-bois-racism-data-paris-african-americans-jobs

45 Eric J. Evans, *Thatcher and Thatcherism*, London: Psychology Press, 2004, p.30.

46 Ian Simpson, *Public Confidence in Official Statistics – 2016*, London: NatCEN social research, 2017, https://natcen.ac.uk/media/1361381/natcen_public-confidence-in-official-statistics_web_v2.pdf

47 The Cabinet Office, *Review of Pre-Release Access to Official Statistics*, https://assets.publishing.service.gov.uk/government/uploads/system/uploads/attachment_data/file/62084/pre-release-stats.pdf

48 Mike Bird, 'Lucky, Good or Tipped Off? The Curious Case of Government Data and the Pound', *Wall Street Journal*, 26 April 2017; and 'New Data Suggest U.K. Government Figures Are Getting Released Early', *Wall Street Journal*, 13 March 2017.

Rule Nine: Remember that misinformation can be beautiful too

1 For more information about the life and statistical contribution of Florence Nightingale, see Mark Bostridge, *Florence Nightingale: The Woman and Her Legend*, London: Penguin, 2009; Lynn McDonald (ed.), *The Collected Works of Florence Nightingale*, Waterloo, Ont: Wilfrid Laurier University Press, 2009-10, and 'Florence Nightingale: Passionate Statistician', *Journal of*

Holistic Nursing, 28(1), March 2010; Hugh Small, 'Did Nightingale's "Rose Diagram" save millions of lives?', seminar paper, Royal Statistical Society, 7 October 2010; Cohen, I. Bernard. 'Florence Nightingale', *Scientific American*, 250(3), 1984, 128–37, www.jstor.org/stable/24969329, accessed 13 Mar. 2020; Eileen Magnello, 'Florence Nightingale: A Victorian Statistician', *Mathematics in School*, May 2010, and 'The statistical thinking and ideas of Florence Nightingale and Victorian politicians', *Radical Statistics*, 102.

2 Draft from John Sutherland (presumed on behalf of Florence Nightingale) to William Farr, March 1861.

3 These quotes from Nightingale are in Marion Diamond and Mervyn Stone, 'Nightingale on Quetelet', *Journal of the Royal Statistical Society*, 1, 1981, 66–79.

4 Alberto Cairo, *The Functional Art*, Berkeley, CA: Peachpit Press, 2013.

5 Robert Venturi, Denise Scott Brown, Steven Izenour, *Learning from Las Vegas: The Forgotten Symbolism of Architectural Form*, Cambridge, MA: MIT Press, 1977; see also https://99percentinvisible.org/article/lessons-sin-city-architecture-ducks-versus-decorated-sheds/; and Edward Tufte, *The Visual Display of Quantitative Information*, Cheshire, CT: Graphics Press, 1983, 2001, pp.106–121.

6 Scott Bateman, Regan L. Mandryk, Carl Gutwin, Aaron Genest, David McDine, Christopher Brooks, 'Useful Junk? The Effects of Visual Embellishment on Comprehension and Memorability of Charts', *ACM Conference on Human Factors in Computing Systems (CHI)*, 2010.

7 Linda Rodriguez McRobbie, 'When the British wanted to camouflage their warships, they made them dazzle', *Smithsonian Magazine*, 7 April 2016, https://www.smithsonianmag.com/history/when-british-wanted-camouflage-their-warships-they-made-them-dazzle-180958657/

8 David McCandless, *Debtris US*, 30 December 2010, https://www.youtube.com/watch?v=K7Pahd2X-eE

9 https://informationisbeautiful.net/visualizations/the-billion-pound-o-gram

10 Brian Brettschneider, 'Lessons from posting a fake map', Forbes.com, 23 November 2018, https://www.forbes.com/sites/brianbrettschneider/2018/11/23/lessons-from-posting-a-fake-map/#5138b31959ec

11 Florence Nightingale, 'Notes on the Health of the British Army', quoted in Lynn McDonald (ed.), *The Collected Works of Florence Nightingale*, vol. 14, p.37.

12 McDonald (ed.), *The Collected Works of Florence Nightingale*, vol. 14, p.551.

13 Letter from Florence Nightingale to Sidney Herbert, 19 August 1857.

14 Alberto Cairo, *How Charts Lie*, New York: W. W. Norton, 2019, p.47.

15 William Cleveland, *The Elements of Graphing Data*, Wadsworth: Monterey, 1994; Gene Zelazny, *Say it with Charts*, New York: McGraw-Hill, 1985; Naomi Robbins, *Creating More Effective Graphs*, New Jersey: Wiley, 2005.

16 Edward Tufte, *Envisioning Information*, Cheshire CT: Graphics Press, 1990.

17 Larry Buchanan, 'Idea of the Week: Inequality and New York's Subway', *New Yorker*, 15 April 2013, https://www.newyorker.com/news/news-desk/idea-of-the-week-inequality-and-new-yorks-subway

18 Simon Scarr, 'Iraq's Bloody Toll', *South China Morning Post*, https://www.scmp.com/infographics/article/1284683/iraqs-bloody-toll
19 Andy Cotgreave, 'Lies, Damned Lies and Statistics', *InfoWorld*, https://www.infoworld.com/article/3088166/why-how-to-lie-with-statistics-did-us-a-disservice.html
20 Letter from William Farr to Florence Nightingale, 24 November 1863, quoted in John M. Eyler, *Victorian Social Medicine: The Ideas and Methods of William Farr*, London: Johns Hopkins Press, 1979, p.175.
21 https://www.sciencemuseum.org.uk/objects-and-stories/florence-nightingale-pioneer-statistician

Rule Ten: Keep an open mind

1 Leon Festinger, Henry Riecken and Stanley Schachter, *When Prophecy Fails*, New York: Harper-Torchbooks, 1956.
2 Walter A. Friedman, *Fortune Tellers: The Story of America's First Economic Forecasters*, Princeton: Princeton University Press, 2013; and Sylvia Nasar, *Grand Pursuit*, London: Fourth Estate, 2011.
3 Friedman, *Fortune Tellers*.
4 Irving Fisher, *How to Live*, New York: Funk and Wagnalls, 21st edition, 1946.
5 Mark Thornton, *The Economics of Prohibition*, Salt Lake City: University of Utah Press, 1991.
6 Esther Ingliss-Arkell, 'Did a case of scientific misconduct win the Nobel prize for physics?', https://io9.gizmodo.com/did-a-case-of-scientific-misconduct-win-the-nobel-prize-1565949589
7 Richard Feynman, 'Cargo Cult Science', speech at Caltech, 1974: http://calteches.library.caltech.edu/51/2/CargoCult.htm
8 M. Henrion and B. Fischhoff, 'Assessing Uncertainty in Physical Constants', *American Journal of Physics*, 54, 1986, 791–8, https://doi.org/10.1119/1.14447
9 Author interview with Jonas Olofsson, 22 January 2020.
10 T. C. Brock and J. L. Balloun, 'Behavioral receptivity to dissonant information', *Journal of Personality and Social Psychology*, 6(4, Pt.1), 1967, 413–28, https://doi.org/10.1037/h0021225
11 B. Fischhoff and R. Beyth, '"I knew it would happen": Remembered probabilities of once-future things', *Organizational Behavior & Human Performance*, 13(1), 1975, 1–16, https://doi.org/10.1016/0030-5073(75)90002-1
12 Philip Tetlock, *Expert Political Judgement*, Princeton: Princeton University Press, 2005; Philip Tetlock and Dan Gardner, *Superforecasting: The Art and Science of Prediction*, New York: Crown, 2015, p.184.
13 Welton Chang, Eva Chen, Barbara Mellers, Philip Tetlock, 'Developing expert political judgment: The impact of training and practice on judgmental accuracy in geopolitical forecasting tournaments', *Judgment and Decision Making*, 11(5), September 2016, 509–26.
14 Tetlock and Gardner, *Superforecasting*, p.127.

15 Nasar, *Grand Pursuit*; and John Wasik, *Keynes's Way to Wealth*, New York: McGraw-Hill, 2013.

16 Anne Emberton, 'Keynes and the Degas Sale', *History Today*, 46(1), January 1996; Jason Zweig, 'When Keynes Played Art Buyer', *Wall Street Journal*, 30 March 2018; 'The Curious Tale of the Economist and the Cezanne in the Hedge', 3 May 2014, https://www.bbc.co.uk/news/magazine-27226104

17 David Chambers and Elroy Dimson, 'Retrospectives: John Maynard Keynes, Investment Innovator', *Journal of Economic Perspectives*, 27(3), 2013, 213-28, https://doi.org/10.1257/jep.27.3.213

18 M. Deutsch and H. B. Gerard, 'A study of normative and informational social influences upon individual judgment', *Journal of Abnormal and Social Psychology*, 51(3), 1955, 629–36, https://doi.org/10.1037/h0046408

19 Philip Tetlock, Twitter, 6 January 2020, https://twitter.com/PTetlock/status/1214202229156016128.

20 Nasar, *Grand Pursuit*, p.314.

21 Friedman, *Fortune Tellers*.

The Golden Rule: Be curious

1 Orson Welles, remarks to students at the University of California Los Angeles, 1941.

2 Onora O'Neill, Reith Lectures 2002, Lecture 4: 'Trust and transparency', http://downloads.bbc.co.uk/rmhttp/radio4/transcripts/20020427_reith.pdf

3 Dan M. Kahan, David A. Hoffman, Donald Braman, Danieli Evans Peterman and Jeffrey John Rachlinski, '"They Saw a Protest": Cognitive Illiberalism and the Speech-Conduct Distinction', 5 February 2011, Cultural Cognition Project Working Paper no. 63; *Stanford Law Review*, 64, 2012; Temple University Legal Studies Research Paper no. 2011–17, available at: https://ssrn.com/abstract=1755706

4 Dan Kahan, 'Why Smart People Are Vulnerable to Putting Tribe Before Truth', *Scientific American: Observations*, 3 December 2018, https://blogs.scientificamerican.com/observations/why-smart-people-are-vulnerable-to-putting-tribe-before-truth/; Brian Resnick, 'There may be an antidote to politically motivated reasoning. And it's wonderfully simple', Vox.com, 7 February 2017, https://www.vox.com/science-and-health/2017/2/1/14392290/partisan-bias-dan-kahan-curiosity; D. M. Kahan, A. Landrum, K. Carpenter, L. Helft and K. Hall Jamieson, 'Science Curiosity and Political Information Processing', *Political Psychology*, 38, 2017, 179–99, https://doi.org/10.1111/pops.12396

5 Author interview with Dan Kahan, 24 November 2017.

6 J. Kaplan, S. Gimbel and S. Harris, 'Neural correlates of maintaining one's political beliefs in the face of counterevidence', *Scientific Reports*, 6(39589), 2016, https://doi.org/10.1038/srep39589

7 G. Loewenstein, 'The psychology of curiosity: A review and reinterpretation', *Psychological Bulletin*, 116(1), 1994, 75–98, https://doi.org/10.1037/0033-2909.116.1.75

8 L. Rozenblit and F. Keil, 'The misunderstood limits of folk science: an illusion of explanatory depth', *Cognitive Science*, 26, 2002, 521–62, https://doi.org/ 10.1207/s15516709cog2605_1
9 P. M. Fernbach, T. Rogers, C. R. Fox and S. A. Sloman, 'Political Extremism Is Supported by an Illusion of Understanding', *Psychological Science*, 24(6), 2013, 939–46, https://doi.org/10.1177/0956797612464058
10 Steven Sloman and Philip M. Fernbach, 'Asked to explain, we become less partisan', *New York Times*, 21 October 2012.
11 Michael F. Dahlstrom, 'Storytelling in science', *Proceedings of the National Academy of Sciences*, 111 (Supplement 4), September 2014, 13614–20, https://doi.org/ 10.1073/pnas.1320645111
12 Bruce W. Hardy, Jeffrey A. Gottfried, Kenneth M. Winneg and Kathleen Hall Jamieson, 'Stephen Colbert's Civics Lesson: How Colbert Super PAC Taught Viewers About Campaign Finance, Mass Communication and Society', *Mass Communication and Society* 17(3), 2014, 329–53, https://doi.org/ 10.1080/15205436.2014.891138
13 'The Planet Money T-Shirt': https://www.npr.org/series/262481306/planet-money-t-shirt-project-series?t=1580750014093
14 *Economics: The Profession and the Public*, seminar held at the Treasury in London, 5 May 2017.
15 Quote Investigator: https://quoteinvestigator.com/2015/11/01/cure/
16 'Why is this lying bastard lying to me?' – this sentiment courtesy of the renowned British reporter Louis Herren.

Acknowledgements

It's been nearly fifteen years since Nicola Meyrick emailed me out of the blue to suggest that I might want to present a Radio 4 programme about statistics. Ever since then, I've been part of the *More or Less* family and it's been both a pleasure and a privilege. This book reflects everything I've learned over the years, so Nicola deserves the credit for starting it all.

I'm grateful to everyone at the BBC who has worked researching, producing, reporting and mixing *More or Less*, doing everything in their power to make me sound good. My back-of-the-envelope estimate is that about a hundred people have been part of the team over the years, but in particular Richard Fenton-Smith and Lizzy McNeill worked on the stories about premature births and gun violence that illustrated Chapter Three. I am especially lucky to have worked with my long-suffering editor, Richard Vadon, and a series of superb producers, notably Ruth Alexander, Innes Bowen, Richard Knight, Kate Lamble and Charlotte McDonald. Andrew Dilnot, who co-created *More or Less* with Michael Blastland, was extremely generous when I asked for his advice. He has been so ever since.

Under Hetan Shah's leadership, everyone at the Royal Statistical Society made me feel every inch an honorary statistician. I'm grateful to them all. Two statistical gurus in particular have been so helpful to me in pondering this book: Denise Lievesley and David Spiegelhalter.

David Bodanis, Paul Klemperer and Bill Leigh all made invaluable comments after reading the entire manuscript – truly a selfless act – and Bruno Giussani caught an important error in an early draft. At Pushkin Industries, Julia Barton, Ryan Dilley, Mia Lobel and Jacob Weisberg have been a pleasure to work with – as well as supplying comments on a podcast script that helped make Chapter Ten a lot better. Andrew Wright's detailed and insightful editing improved the book a great deal, as it has done many times before; he's a star and a true friend.

Thanks to every scholar and writer whose work I have relied on either through interviews, emails or by consulting their writing, in particular: Anjana Ahuja, Michael Blastland, Alberto Cairo, Andy Cotgreave, Kate Crawford, Kenn Cukier, Andrew Dilnot, Anne Emberton, Baruch Fischhoff, Walter Friedman, Hannah Fry, Kaiser Fung, Dan Gardner, Andrew Gelman, Ben Goldacre, Rebecca Goldin, David Hand, Dan Kahan, Daniel Kahneman, Eileen Magnello, Viktor Mayer-Schönberger, Lynn McDonald, David McRaney, Barbara Mellers, Errol Morris, Will Moy, Terry Murray, Sylvia Nasar, Cathy O'Neil, Onora O'Neill, Caroline Criado Perez, Robert Proctor, Jason Reifler, Alex Reinhart, Anna Rosling Rönnlund, Max Roser, Hans Rosling, Benjamin Scheibehenne, Janelle Shane, Hugh Small, Lucy Smith, Philip Tetlock, Edward Tufte, Patrick Wolfe, David Wootton, Frank Wynne, Ed Yong and Jason Zweig.

At Little, Brown, Tim Whiting and Nithya Rae have been models of patience as I embarked on an extended coronavirus rewrite. Dan Balado and Holly Harley made invaluable edits, as did my editor at Riverhead Books in the US, Jake Morrissey. Thanks of course to the implacably excellent Sally

Holloway, to Zoe Pagnamenta, and everyone at Felicity Bryan Associates.

I continue to value the support and indulgence of my editors at the *Financial Times*, in particular Alice Fishburn, Brooke Masters and Alec Russell. Loyal FT readers will see that some of the ideas in this book were first explored in my writing for the newspaper. I love the FT and am so glad to be part of it.

Thank you to my children, Stella, Africa and Herbie, just for being you. And to Fran Monks; I'm not even going to try to count the ways I am grateful – they would fill another book.

Credits

Introduction

Umberto Eco, *Serendipities: Language and Lunacy*, London: Orion, 1998.

Rule One

The Empire Strikes Back (1980); also known as *Star Wars: Episode V*; screenplay by Leigh Brackett and Lawrence Kasdan.

Rule Two

Muhammad Yunus in an interview by Steven Covey, socialbusinesspedia.com/wiki/details/248.

Rule Three

Douglas Adams, *Hitchhiker's Guide to the Galaxy*, London: Pan Books, 1979.

Rule Four

Terry Pratchett, *Reaper Man*, London: Victor Gollancz, 1991.

Rule Five

Alan Moore, *Watchmen*, New York: DC Comics, 1986.

Rule Six

Anna Powell-Smith, MissingNumbers.org.

Rule Seven

2001: A Space Odyssey (1968); screenplay by Stanley Kubrick and Arthur C. Clarke.

Rule Eight

Hans Rosling, translation of a Danish TV interview, thelocal.se/20150905/hans-rosling-you-cant-trust-the-media

Rule Nine

Michael Blastland, personal correspondence, 13 May 2013.

Rule Ten

Leon Festinger, Henry Riecken and Stanley Schachter, *When Prophecy Fails*, New York: Harper-Torchbooks, 1956.

The Golden Rule

Orson Welles, in remarks to students at the University of California Los Angeles, 1941.

Index

About the Author

About the author

Tim Harford is a senior columnist for the *Financial Times*, and the presenter of *Cautionary Tales* and Radio 4's *More or Less*.

He is an honorary fellow of the Royal Statistical Society, a member of Nuffield College, Oxford, and the winner of numerous awards for economic and statistical journalism.

In 2019 he was made an OBE 'for services to improving economic understanding'. Tim lives in Oxford with his wife and three children. His other books include *The Undercover Economist, Adapt, Messy, Fifty Things that Made the Modern Economy* and *The Next Fifty Things that Made the Modern Economy.*